AF713205

Jahan Marcu, Andrew M. Peterson (Eds.)
Regulated *Cannabis* and Hemp Market Navigation

Cannabis Innovations

Series Editors
Jahan Marcu and Andrew M. Peterson

Volume 1

Regulated *Cannabis* and Hemp Market Navigation

Business, Public Health, and Research Challenges

Edited by
Jahan Marcu and Andrew M. Peterson

DE GRUYTER

Editors
Dr. Jahan Marcu
3R 285 Pulaski St.
Brooklyn, NY 11206
United States of America
MarcuEnterprises@gmail.com

Andrew M. Peterson
623 Kingsley Drive
Ventnor City, NJ 08406
United States of America
apeterson@sju.edu

ISBN 978-3-11-067742-3
e-ISBN (PDF) 978-3-11-067750-8
e-ISBN (EPUB) 978-3-11-067760-7

Library of Congress Control Number: 2024933577

Bibliographic information published by the Deutsche Nationalbibliothek
The Deutsche Nationalbibliothek lists this publication in the Deutsche Nationalbibliografie; detailed bibliographic data are available on the internet at http://dnb.dnb.de.

Cover image: janiecbros/E+/Getty Images
Typesetting: Integra Software Services Pvt. Ltd.
Printing and binding: CPI books GmbH, Leck

www.degruyter.com

Contents

Preface

Andrew and I met through the University of the Sciences during the pandemic.

As we navigated the uncertain waters of the pandemic, where remote interaction was the norm, our paths converged. I, fresh into an adjunct faculty role, was immersing myself in *Cannabis* studies, while Andrew had just completed his fifth edition of a significant publication. Our mutual introduction sparked a journey of collaboration and exploration in the realm of *Cannabis* literature.

Our initial foray into pitching ideas to publishers was met with a series of rejections. These setbacks, however, proved invaluable. They prompted us to reassess our goals and aspirations. Did we really want to add to the pile of basic, often outdated *Cannabis* textbooks? This introspection led us to reshape our concept into something more ambitious, more in line with our true aspirations.

We envisioned a book that broke the mold – chapters penned by diverse professionals actively shaping the *Cannabis* industry. We sought voices from a spectrum of disciplines: regulatory consultants, lawyers, pharmacologists, industry operators, and university academics. This diversity in perspective and expertise is something we found conspicuously absent in existing *Cannabis* literature.

Our shared passion has always been to explore various perspectives and writing styles. This book is the embodiment of that passion, a result of our years of collaborative effort. While the publishing world seemed fixated on beginner-level texts, we aspired for something more substantial. We understood the limitations of extensive textbooks – often outdated before they even reach readers. Instead, we pitched a book that aligned with our dream: a compilation of advanced, complex ideas in *Cannabis*, targeting readers who already possess a foundational understanding.

This is not just a book; it's a multidisciplinary and multimedia endeavor, crafted for swift consumption and rapid production. If this resonates with you, we warmly invite you to consider contributing your own insights for future editions.

If you're reading this, chances are you too are intrigued by the potential for innovation in *Cannabis*. Among our contributors are long-standing friends, like Ethan Russo, whom one of us met during undergraduate days, and others whom we've admired and now had the pleasure of collaborating with through this project.

What lies within these pages is more than just text; it's a dialogue that we're excited to share with you, and one that we hope you'll join.

https://doi.org/10.1515/9783110677508-203

List of contributing authors

Mackenie Slade, MPH
Cannabis Public Policy Consulting
490-B Boston Post Road
Sudbury, MA 01776
mslade@cannabispublicpolicyconsulting.com

Jahan Marcu, PhD
Marcu Enterprises
347 8th St Suite 1
Jersey City, NJ 07302
marcuenterprises@gmail.com

Jason Schechter, PhD
Cortical Systematics LLC
PO BOX 69772
Oro Valley, AZ 85737-9998
jbs@corticalsystematics.net

Sandra Carrillo, MD, MHSM
Colombian Medical *Cannabis* Association
Torre de las Americas
Medellín, Colombia
drsandracarrillo@gmail.com

Thomas A. Trite, BSPharm, RPh, FASCP
PA Options for Wellness, Inc.
4711 Queen Ave
Harrisburg, PA 17109
ttrite@paofw.com

Tzuo-Zen Lee, ChE, BSPharm, RPh
PA Options for Wellness, Inc.
4711 Queen Ave
Harrisburg, PA 17109
tlee@paofw.com

Andrew M. Peterson, PharmD, PhD, FCPP
Philadelphia College of Pharmac
Saint Joseph's University
600 South 43rd Street
Philadelphia, PA 19104
apeterson@sju.edu

Sarah Trent, Esq.
Valley Wellness
407 US 202
Raritan, NJ 08869
sarah.trent@valleywellnessnj.com

Seth Tipton, Esq.
Florio Perrucci Steinhardt Cappelli & Tipton LLC
91 Larry Holmes Drive
Suite 200
Easton. PA 18042
stipton@floriolaw.com

Ruth Charbonneau, JD, RN
College of Health Professions (JCHP)
Thomas Jefferson University
901 Walnut Street
Philadelphia, PA 19107
Ruth.Charbonneau@jefferson.edu

Kevin H. Fortin, Esq., JD, BSME
Fortin Patent Firm
1400 Shoshoni Camp Road
Rollinsville, CO 80474
kevin@fortinpatent.com

Miles W. Tuttle, MS, MBA, MPA
William S. Richardson School of Law
University of Hawai'i
2515 Dole Street
Honolulu, HI 96822-2350
Milesw@hawaii.edu

Teresa Simon, MPH
Physicians Research Center
601 Route 37 East
Toms River, NJ 08755
prctrials@gmail.com

https://doi.org/10.1515/9783110677508-204

Mackenzie Slade

Innovating in *Cannabis* policy

Introduction

Since 2012, adult use *Cannabis* legalization has arisen as one of the America's greatest social experiments. Serving as the catalyst to a newfound awareness of the limitations of federalism, the state-confined legalization of *Cannabis* is an innovation in public policy. Before beginning to identify avenues that further innovate *Cannabis* policy, we must start with this premise: that a regulatory infrastructure tailored to create a limited yet inclusive economy with a federally illicit controlled substance is intrinsically innovative. As with all innovative policies, diffusion across jurisdictions has become rampant. In the 10-year period from the "birth year" of *Cannabis* legalization for adult consumers to the year of preparing this chapter, 22 states have legalized *Cannabis* for adult use. In this same timeframe, 23 states legalized *Cannabis* for medical consumers, leading to a total of 38 states with a form of legalized *Cannabis*. Many of these states are overlapping, suggesting that the liberalization of *Cannabis* policies within jurisdictions has also been seen with urgency [1].

In many ways, this accelerated proliferation of *Cannabis* legalization across U.S. states has outpaced our ability to identify avenues for originality. For better or for worse, states have been largely imitating each other's policies in an effort to prioritize the urgency of consumer demand rather than innovation. This statement should not be misconstrued as a criticism of state programs or lawmakers. Regulators walk a metaphorical tightrope of enforcing policies that protect public health and society yet support the *Cannabis* industry. This balancing act is only further complicated by the complexities of establishing the *Cannabis* industry. *Cannabis* markets have a harsh sensitivity to path dependence, thriving substitute markets, and unprecedented market conditions. States that have experienced hurdles resulting in delayed implementation timelines have struggled to transition consumers over from well-established illicit and substitute alternative cannabinoid markets, even years after launch. States like New York, for example, have found themselves in a cyclic battle of shutting down illicit storefronts while simultaneously trying to find outlets for their own regulated *Cannabis* supply as *Cannabis* cannot be transported outside of the state, nor outside of the limited regulated system [2]. When taking all of this into consideration, in addition to the political discourse surrounding the topic, it becomes clear that the intricacies of *Cannabis* policy are in a league of their own.

Any attempt to abandon the current *Cannabis* policy paradigm and employ a novel change comes with innate risks. Many of these risks can offset the regulatory balancing

Mackenzie Slade, *Cannabis* Public Policy Consulting

https://doi.org/10.1515/9783110677508-001

act, resulting in damaging public health or industry impacts. Importantly, a risk to the industry is a risk to business owners, many of whom have disproportionately suffered from *Cannabis* criminalization and have put their life savings into the hope of finally being able to operate in a regulated environment – a hope worthy of support.

It is because of this challenging risk/reward ratio that **innovating in *Cannabis* policy must be viewed as a practice, not an outcome.** The slew of problems that exist and would benefit from innovation is no secret. However, the ways to address them without teetering off the tightrope have yet to be fully discovered. For regulators and lawmakers, the focus of innovation should center less on the policy to create (the "what") and more on the process by which the policy can be created (the "how").

In focusing on the "how," we reject the idea that innovation must come in the form of a singular, disruptive, technology-enabled proposal. How innovation occurs inside the *Cannabis industry* is different than how innovation occurs in the context of *Cannabis policymaking*. *Cannabis* policy is not a business, and the regulator's role in governing *Cannabis* is not to turn a profit for shareholders but to ensure the *Cannabis* economy meets the demands of all stakeholders. The practice of innovation in *Cannabis* policy will more closely resemble other areas of policy, bringing the field away from its current infancy into an era of modernity.

The remainder of this chapter will illustrate the three directions in which the public sector must look to innovate *Cannabis* policy: *outward*, *forward*, and *inward*. These three reflective positions and actions taken within them should identify areas of improvement, implement innovation, and demonstrate improved value across *Cannabis* policies.

Part I. Looking outward

Most experiments, whether in a lab or real-world setting, follow the scientific method. They begin with a simple observation, one that ultimately leads to a hypothesis. That hypothesis is then tested and analyzed until conclusions can be drawn. **If *Cannabis* legalization is one of America's greatest social experiments, where is the analysis?**

While there is no shortage of intermittent studies measuring the impacts of *Cannabis* legalization on hyper-specific public health outcomes, there are very few governments conducting systematic evaluations on the successes or failures of state *Cannabis* programs in a manner that enables continuous improvement in policy development. In order to innovate *Cannabis* policy, we first must evaluate it. As the saying goes, you can't improve what you don't measure.

This may seem rather obvious in theory, but it is more challenging in practice, as there is no standard evaluation framework or validated performance measures within programs for regulators to monitor and respond to. *Cannabis* policy research underpins innovation; the absence of research-based frameworks has stifled innovation as well as our ability to be pragmatic.

While the field works toward consensus on what metrics point to the ongoing successes or failures of a market, regulators can begin their practice of innovation with testing the most foundational hypothesis of *Cannabis* policy: regulating *Cannabis* will facilitate the eradication of an illicit market. As you may have guessed, the only way to test this hypothesis is by *Looking Outward*, toward consumers and the drivers of their demand.

Importantly, in order for the practice of innovation to be effective, all evaluations must inform learning. This means that data should not be collected with the sheer goal of identifying and cataloging numbers year over year. This is not to say that looking at trends cannot be helpful in the practice of innovating policy; it can be. However, research should prioritize informing policymaking decisions in real time. **Innovation in policy happens when policymakers can move from being reactive to being proactive.** When testing our foundational hypothesis, the following four questions can quantify market performance and inform learning:

1. **Who**: What is the prevalence of past-year, past-month, and daily *Cannabis* consumption among the population, and what is the geographic and demographic breakdown of this prevalence?
2. **What**: What proportion of these consumers are participating in the regulated market as opposed to illicit or other substitute markets?
3. **How**: What is the average volume of *Cannabis* that consumers are accessing from each *Cannabis* source they have identified?
4. **Why**: What are the factors influencing consumer participation in each *Cannabis* source?

In answering these four basic questions through empirical surveillance such as the Regulatory Determinants of *Cannabis* Outcomes Survey [3], policymakers can rapidly assess the performance of their regulated market against illicit markets and, most importantly, identify where shortcomings may exist. If done correctly, the answers to these initial questions should lead to a sequence of others, initiating an investigation that results in policy innovation.

Maine's Office of *Cannabis* Policy (OCP) provides an excellent case study in this. In 2021, OCP commissioned empirical surveillance on market dynamics in an effort to understand how the first year of an active adult use market performed. Research found that Maine's regulated *Cannabis* market was incredibly successful at the end of year 1, resulting in a higher regulated market capture than expected when accounting for the amount of time the market had been active [4]. Figure 1 below shows a recently updated national analysis reflecting similar findings over two years after Maine's market launch, with a predictive performance line demonstrating expectations of legal market capture. Importantly, Maine's analysis also showed that consumers prioritized their source of *Cannabis* based on price, a common priority factor in *Cannabis* purchasing, as *Cannabis* consumers are a price-sensitive population [5].

While this finding is not exactly novel, the research answering questions #1 and #4 in the above list pointed to something of interest: there was a subset of past-month consumers living in specific zip codes without accessible adult use stores who were willing to pay a higher monetary amount for a gram of *Cannabis* than the average Maine consumer. These data points coupled together suggested that should adult use *Cannabis* become available in these zip codes, consumers may transition away from illicit sources and toward an adult use market, pushing that successful regulated market capture even further. This research led the regulators of Maine into a series of follow-up questions, as any good policy evaluation should.

The first in the series of investigative questioning was identifying why these zip codes do not have adult use stores. In time, it became apparent that the zip codes in question had not opted in for adult use *Cannabis* retail within their municipality primarily because there was limited incentive to. Unlike in other states, municipalities in Maine cannot levy a local tax on *Cannabis*, making the administration of a local program costly. Looking outward, regulators saw this rationale as an opportunity to innovate. OCP began by requesting municipal revenue sharing at 5% from the state's 15.5% taxation on *Cannabis*. This original request was denied by the legislature, but OCP continued to negotiate [7]. Importantly, the research found that illicit markets had associations to worsening public health outcomes such as driving under the influence of *Cannabis*, tobacco use, and problematic *Cannabis* use. Meanwhile, the regulated market had no associations with these poor health outcomes.

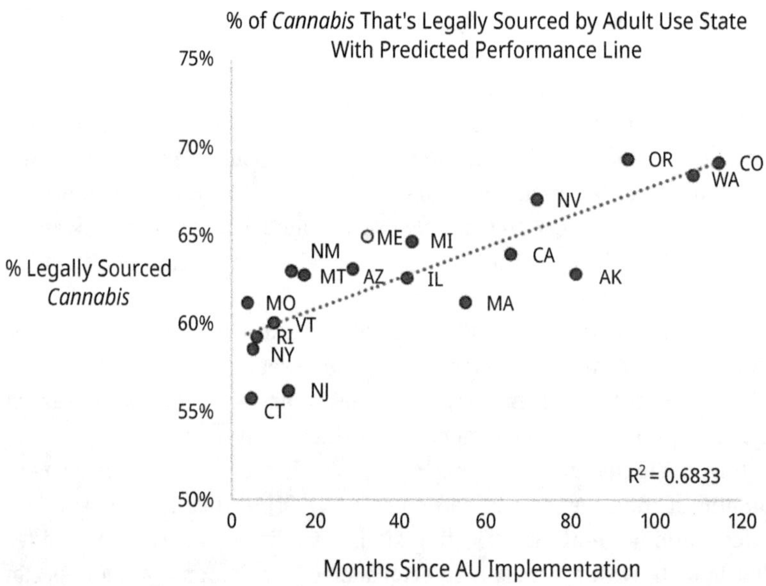

Figure 1: Estimated Legally/Illicitly Sourced and Time Since Legalization by State (Regulatory Determinants of *Cannabis* Outcomes Survey, Data Collected and Analyzed June 2023) [6].

Using this groundbreaking finding, OCP was able to request a bill providing municipalities the opportunity to recover local *Cannabis* program administration costs and legal fees up to $20,000 as an incentive to opt into adult use cultivation and retail stores through their Public Health and Safety Fund. The funds for this endeavor were carved out at 25% [8]. Carve-out funds are not common, and this request did not come without justification from the regulators, wherein being armed with data was proven the best defense. By looking outward through their program evaluation, Maine was able to reimburse more than a dozen municipalities for their decision to opt in to regulated adult use *Cannabis* sales. This fund may very well be the impetus for additional municipalities to opt in and expand market capture in the following years.

It is worthy of note that the innovation observed in this case study could have occurred regardless of the data points provided by consumers. Maine may have been able to host a series of stakeholder engagements and identify a need for a municipal fund prior to launching their market. However, what the policy evaluation offered was clarity on where to begin. The empirical data demonstrated that widespread municipality opt-in would be beneficial for continued success of the regulated market and identified the exact zip codes where expansion would be *most* beneficial.

Regulators are constantly surrounded by the opinions and subjective expertise of the public. Data, when collected with validated methods, is objective. It cancels out the noise and provides direction. When regulators look outward, they are rewarded with efficiencies in policy decision-making. By definition, this is innovating in *Cannabis* policy.

Part II. Looking forward

In many arguments within *Cannabis* policymaking, it is commonplace to hear cries for regulating *Cannabis* like it is alcohol or tobacco. While these arguments may have been more familiar in the early days of legalization, prior to policy imitation becoming well established, they still resonate today. **Ten years after the first successful ballot measure for adult use *Cannabis*, it is time that we regulate *Cannabis* like it is . . . *Cannabis*.** We do this through what is commonly referred to as "experimental policymaking."

Experimental policymaking focuses on testing the effects of a policy decision in real-world or simulated settings and evaluating the outcomes prior to execution [9]. In the world of data science, we call this predictive modeling. Recall from the earlier section that looking at trends can be helpful primarily outside of real-time decision-making. This is where predictive modeling comes in, analyzing patterns and trends within a data set to create sophisticated mathematical models that can predict future outcomes under certain market conditions. *Looking Forward* must come immediately after *Looking Outward*, as these data sets are inextricably linked.

While predictive modeling is the twenty-first-century advancement of many modern industries, the utility of predictive modeling for policy planning is best observed in the healthcare industry. Take the COVID-19 pandemic, for example. Upon discovering the traceability of the new pathogen spreading across the globe, epidemiologists began documenting the epidemic's growth patterns across temporal and social factors. These complex data sets essentially quantified the reproduction and transmission of the pathogen in geographies across time, identifying the size of outbreaks, the maximum period for outbreaks, and subsequent spread times in a predictive manner [10]. These data sets were undeniably helpful in capacity planning within local healthcare systems, prevention messaging, and identifying the appropriate incubation period for each variant. Moreover, these models were also helpful in guiding governments to execute policies like "Safer at Home" mandates given a prediction of increased spread or peak morbidity [11].

While policies surrounding COVID-19 may sound like an extreme deviation from the topic of *Cannabis* policy, the application is relevant. Predictive modeling is being utilized to inform policies across the globe, some policy decisions as high priority as preventing mortality. Surely, it can and should be used to inform something as limited as the current U.S. *Cannabis* market, principally if we use data to test our policies prior to executing them.

The introduction of this chapter briefly pointed to the *Cannabis* industry's harsh sensitivity to path dependence. In economics and public policy, path dependence is best described as the processes by which past events or decisions constrain later events or decisions [12]. In terms of *Cannabis* policymaking, if we select and execute an arbitrary policy that is so burdensome that it is nearly prohibitive to the industry's success, it may become impractical to reverse or improve upon the original outcomes given transitory conditions. History matters, and, as such, experimental policymaking should be prioritized.

In 2022, the State of Maryland saw experimental policymaking, or predictive modeling, as an opportunity to inform the development of their adult use *Cannabis* statute. The Maryland State General Assembly commissioned the same empirical surveillance discussed in the prior section (*Looking Outward*) with a focus on identifying data points that embody future behaviors of consumer participation within the market (*Looking Forward*).

Combining Maryland consumer data with a sharp focus on future market participation and associated behavioral elements, as well as historical supply and sales data from other states, the *Cannabis* Policy Simulation Lab, a product designed for experimental policymaking in *Cannabis*, was able to predict future regulated market performance based on variations of vital policy factors: the rate of taxation, the amount of adult use retail stores available, and the amount of time that the market has been open [13]. Importantly, these three policy components are confounding variables, meaning they interact with each other greatly when influencing or producing an outcome. In experimental policymaking, simulations should account for variations in multiple variables so as to identify the most optimal outcome possible with multiple policy levers.

What Maryland found through their experimental policymaking research led to the inclusion of two foundational elements in their adult use *Cannabis* statute: taxation and capitation on retail licensing. Using a sophisticated matrixed approach as shown in Tables 1–3, researchers tested the amount of regulated market capture over the course of five years under tax rates at 10, 15, 20, 25, and 30%, while also testing the number of dispensaries at 100–250, 260–500, and 500–800. In real-world settings, the variables of taxation, the period of time that regulated *Cannabis* has been accessible to purchase through licensed retail stores (i.e., months of implementation), and the volume of retail stores available all interact in influencing regulated *Cannabis* sales. The following tables attempt to show the outcome of those three variables interacting when all three policy variables are increased or decreased in an attempt to find the most optimal trade-off point for each policy decision with the goal of having the lowest percent of illicit *Cannabis* possible. For the State of Maryland, the simulations of these policies together showed that the most optimal tax rate for regulated market capture was not more than 15%, and with no fewer than 300 dispensaries [14].

Table 1: Maryland Predicted % of Illicit *Cannabis* Between Months 1 and 12 After Adult Use Implementation.

Number of dispensaries	Effective tax rate	Month of implementation		
		1–4	5–8	9–12
100–250	**10%**	43%	42%	41%
	15%	45%	44%	43%
	20%	47%	46%	46%
	25%	49%	49%	48%
	30%	52%	51%	50%
260–500	**10%**	38%	37%	36%
	15%	40%	39%	38%
	20%	42%	41%	41%
	25%	44%	44%	43%
	30%	47%	46%	45%
500–800	**10%**	33%	32%	31%
	15%	35%	34%	33%
	20%	37%	36%	36%
	25%	39%	39%	38%
	30%	42%	41%	40%

Table 2: Maryland Predicted % of Illicit *Cannabis* Between Months 25 and 36 After Adult Use Implementation.

Number of dispensaries	Effective tax rate	Month of implementation		
		25–28	29–32	33–36
100–250	**10%**	38%	37%	36%
	15%	40%	39%	38%
	20%	42%	42%	41%
	25%	45%	44%	43%
	30%	47%	46%	45%
260–500	**10%**	33%	32%	31%
	15%	35%	34%	33%
	20%	37%	37%	36%
	25%	40%	39%	38%
	30%	42%	41%	40%
500–800	**10%**	28%	27%	26%
	15%	30%	29%	28%
	20%	32%	32%	31%
	25%	35%	34%	33%
	30%	37%	36%	35%

Table 3: Maryland Predicted % of Illicit *Cannabis* Between Months 49 and 60 After Adult Use Implementation.

Number of dispensaries	Effective tax rate	Month of implementation		
		49–52	53–56	57–60
100–250	**10%**	33%	32%	31%
	15%	35%	34%	34%
	20%	38%	37%	36%
	25%	40%	39%	38%
	30%	42%	41%	41%
260–500	**10%**	28%	27%	26%
	15%	30%	29%	29%
	20%	33%	32%	31%
	25%	35%	34%	33%
	30%	37%	36%	36%
500–800	**10%**	23%	22%	21%
	15%	25%	24%	24%
	20%	28%	27%	26%
	25%	30%	29%	28%
	30%	32%	31%	31%

In addition to those two simulated policies, the answers to question #4 proposed in the *Looking Outward* section indicated something unique. While *Cannabis* consumers did not prioritize delivery as a factor in purchasing, they did indicate that they would utilize this source specifically, leading to a predicted 13% increase in regulated sales should Maryland allow for delivery.

At the time of this experimental policymaking project, the status quo for launching an adult use *Cannabis* market was transitioning away from waiting until all or a majority of new licensees were prepared to opening the market with existing medical *Cannabis* licensees and having new licensees progressively enter the market. The reason for this approach is to better meet consumer demand with urgency and prevent the illicit market from thriving while the new industry rolls out, as getting a retail store licensed and open with regulated products can take over a year. In addition to this trend, there is a continued struggle to improve the ability of social equity licensees to capture and retain market share. When states launch with medical *Cannabis* licenses, the first-to-market strategy is quickly absorbed. The predictive data point of delivery increasing regulated sales by 13% posed an opportunity to address the ongoing issue of urgency while supporting social equity businesses.

The experimental policymaking project suggested an innovative opportunity to start the market with current medical *Cannabis* licensees while also allowing for social equity licensees to operate a delivery store without the requirement or necessity of a brick-and-mortar location. This model would mirror an Amazon-style model, having lower start-up costs and the ability to begin operating quickly. This would effectively allow a delivery licensee to obtain regulated product quickly from current medical *Cannabis* licensees and to share in the first-to-market strategy.

The experimental policymaking project led to innovation in three foundational components of *Cannabis* legislation. The optimal taxation rates, amount of retail licenses, and the type of delivery model were all incorporated into the bill's initial language, making it arguably the most data-informed and innovative statute introduced in *Cannabis* history. The State of Maryland originally incorporated these components in an effort to get its program right from the start. However, experimental policymaking can happen at any point across the regulatory continuum. States with mature markets can easily test the effects of a policy change prior to implementation. The impact of delivery on regulated sales could serve as an excellent example for those that still prohibit this activity. As Maryland has demonstrated, policymakers can reduce long-term risk, increase efficiencies, and even identify areas of opportunity for originality by looking forward.

Part III. Looking inward

The *Cannabis* industry follows the microeconomic principles that every other industry must face, including the law of supply and demand. *Cannabis*, a consumer good, can experience supply outpacing demand, and price plummeting as a result. *Looking Outward* and *Looking Forward* focus primarily on demand, using consumer behavior data to identify areas of innovation within *Cannabis* policymaking. Looking at demand has served as an avenue for finding opportunities as opposed to problems. However, demand is only one variable in this equation, and identifying and addressing opportunities does not negate the need to address and mitigate problems.

At the time of writing this chapter, overproduction from mature *Cannabis* markets serves as a cautionary tale for those launching new *Cannabis* markets. States such as Colorado and Michigan have experienced overproduction and plummeting prices to the point of business owners inching closer to failure [15]. The industry members from these states have called for their regulators to help mitigate this problem, but for some, the issue is already apparent [16]. This is a perfect example of a missed opportunity for regulators to *Look Inward* at the data they have readily available. Recall, innovation happens when policymakers can move from being reactive to being proactive.

Each *Cannabis* program has critical policy elements that make up its foundation, one vital element being product traceability systems. Traceability within each state *Cannabis* program, also known as Track-and-Trace, is vital for monitoring how *Cannabis* product supply is moving through the supply chain from the point of cultivation to the point of sale. By far, Track-and-Trace is the most exhaustive data system in all of *Cannabis*. Importantly, a form of these systems is statutorily required for nearly every *Cannabis* program in the United States. These systems capture millions of useful data points weekly. **The *Cannabis* field is not devoid of data; it is devoid of efforts to make existing data truly meaningful.** By looking inward at the data that all *Cannabis* programs are mandated to collect, regulators can spot problems in real time and become innovative.

As part of standard *Cannabis* regulatory compliance, licensees are required to report their current production inventory in the Track-and-Trace system. This production inventory is tracked by individual plant count, by canopy square footage, or by both. Relatedly, most states place a maximum capacity for production inventory through licensing tiers. For example, a Tier 1 cultivation licensee may cultivate 500 square feet of canopy, whereas a Tier 4 cultivation licensee may cultivate 20,000 square feet of canopy. Although Tier 4 licensees are permitted to cultivate up to 20,000 square feet of canopy, that maximum capacity is not always reached. Typically, licensees will scale up to meet capacity over time as market capture expands and economies of scale are achieved. This trend is to be expected after launching a market as demand becomes met overtime. Should *Cannabis* programs observe the opposite trend – cultivators beginning to drastically reduce their production inventory and move away from their maximum capacity – an alarm should go off. This trend points to a slowly brewing overproduction problem.

When this trend is observed, a plummet of wholesale prices is likely to coincide, inevitably leading to *Cannabis* products being left without an outlet. This exact observation is best showcased in Oregon, one of the first states to experience a problem with oversupply, resulting in a projected 6.5 years' worth of inventory [17]. A supply and demand report prepared by the Oregon Liquor and *Cannabis* Commission analyzed Track-and-Trace data from 2017 to 2018 demonstrating these trends. This report is an annual legislative mandate and was produced in 2019, months after these trends started to appear and become exacerbated. Instead of being in a position to prevent this colossal market problem, the policymakers in Oregon were now tasked with fixing it. As a result, the Commission established a moratorium on producer licenses, a production management policy utilized in many industries where licenses are required [18]. However, due to a surplus in backlogged applications, production still expanded, and market problems prevailed.

As this story relates to innovation, the critical takeaway is that the data that prompted a reaction to this problem is the same data that could have been used to prevent it. The only difference is, in preventing the problem, policymakers *Look Inward*, monitor data as opposed to aggregate and analyze annually. Real-time monitoring is an example of efficiency, and efficiency is a tool of innovation.

Years after their attempts to correct the market from its 2019 supply glut, policymakers in Oregon learned from this mistake. In 2022, supply data was pointing to a resurgence of the same problem – overproduction. As a result of this market monitoring, the Commission was able to place a moratorium on licensing until March 2024 [19]. While it may be too soon to declare, the timing of this policy is likely to aid in the prevention of the market returning back to 2017 conditions. The annual supply and demand report provided to the legislature this year echoes similar sentiments [20]. When looking inward and viewing the single statutorily required *Cannabis* data system not just as technology for licensee compliance but as a vital information hub, policymakers can make data meaningful, witness problems as they arise, and become efficient.

Interestingly, the 2019 and 2022 policies discussed in this case study are identical in their intent. This fact alone may bring into question how *Looking Inward* may be perceived as a process for innovating in *Cannabis* policy. This brings us back to the original point, that the focus of innovation in *Cannabis* policymaking should center less on the policy to create (the "what") and more on the process by which the policy can be created (the "how"). In 2019, the moratorium was issued too late. The problem had already arisen and had an additional backlog of contributing complications, like pending producer applications and an increasing gray market, that couldn't be overlooked. In 2022, the problem of overproduction was just beginning to rear its head, with a potential market devaluation to follow. The innovation seen here is simply identifying the optimal timing. For a brand-new industry where much of the decision-making is through trial and error, getting something as simple as timing correct can be an example of innovation.

Conclusion

This chapter would be remiss to not make mention of the current climate of *Cannabis*. The "green rush" has seemingly come to an end and businesses across the country have been pushed into turmoil, many closing their operations [21]. At the same time, legalization will undoubtedly prevail. This will include federal involvement, be it through the rescheduling or de-scheduling of *Cannabis* in the Controlled Substance Act, or national *Cannabis* legalization with federal regulations and oversight. Just as policymakers begin to find their footing in their own state-confined markets, the landscape is going to rapidly evolve with interstate commerce. Instead of seeing these changes as a daunting exercise in trial and error, policymakers find themselves in the perfect opportunity to deploy the *Outward, Forward*, and *Inward* practices described in this chapter.

Through appropriate demand surveillance (*Looking Outward*), agencies can have an empirical baseline understanding of their consumer behavior with current market conditions and identify areas ripe for improvement while on the precipice of change. Through experimental policymaking (*Looking Forward*), agencies can make confident predictions about how their consumers may behave in future market conditions, like when interstate sales begin. Finally, by *Looking Inward,* agencies can make the millions of data points they already collect meaningful, and better prepare their market for inevitable national influence.

When interstate commerce is authorized, be it through federal reform, state agreements, or decided through the courts, states will need uniformity in data collection and analyses. Such standardization isn't just vital for research, it could very well be required of evolving state and federal laws. Thankfully, these three practices can be standardized with great ease. Organizations like the *Cannabis* Regulators Association have emerged with this specific intent; to identify best practices and standardize them [22]. As *Cannabis* policy enters its next phase of maturity, innovation isn't just possible, it's assured.

References

[1] Where marijuana is legal in the United States. MJBizDaily. (Accessed February 6, 2022 at https://mjbizdaily.com/map-of-us-marijuana-legalization-by-state/.)

[2] New York State now prepared to shut down illegal cannabis shops. Gov. Hochul says. CBS News, 2023. (Accessed June 22, 2023 at https://www.cbsnews.com/newyork/news/illegal-cannabis-shops-that-continue-sales-will-be-shut-down-hochul-says/.)

[3] Sofis MJ. Regulatory determinants of cannabis outcomes survey. Cannabis Public Policy Consulting, 2022. (Accessed March 3, 2023 at https://www.cannabispublicpolicyconsulting.com/wp-content/uploads/2022/06/Report-4.27-V.2.1.pdf.)

[4] Sofis M, Slade M. Maine Office of Cannabis Policy Cannabis Markets & Associated Outcomes – Survey Findings and Implications. Sudbury, MA: Advocates for Human Potential, Inc, 2022. (Accessed March 3, 2023 at https://www.maine.gov/dafs/ocp/sites/maine.gov.dafs.ocp/files/inline-files/Maine%20OCP%20AHP%20Report%2006-22_0.pdf.).

[5] Amlung M, Reed DD, Morris V, Aston ER, Metrik J, MacKillop J. Price Elasticity of Illegal Versus Legal Cannabis: A Behavioral Economic Substitutability Analysis. Addiction 2019;114:112–118. (Accessed March 3, 2023 at https://doi.org/10.1111/add.14437.).

[6] Sofis MJ. Regulatory determinants of cannabis outcomes survey. Cannabis Public Policy Consulting, 2022. (Accessed June 23, 2023 at https://www.cannabispublicpolicyconsulting.com/regulatory-determinants-of-cannabis-outcomes-survey/)

[7] Wade C. Tax revenue: Maine will share its pot with local governments. The Center Square, 2022. (Accessed March 3, 2023 at https://www.thecentersquare.com/maine/tax-revenue-maine-will-share-its-pot-with-local-governments/article_eb8659be-c639-11ec-8202-7b7642805c4e.html#:~:text=The%20measure%2C%20which%20was%20recently,up%20to%20%2420%2C000%20a%20year.)

[8] An act to require the state to distribute 25 percent of adult use marijuana retail sales and excise tax revenue to generating municipalities. 129th Maine Legislature, HP0260, LD 335, item 1. (Accessed March 3, 2023 at https://legislature.maine.gov/legis/bills/bills_129th/billtexts/HP026001.asp.)

[9] Exploring policy innovation: Tools, techniques + approaches. Toronto, ON, Brookfield Institute for Innovation +Entrepreneurship, 2018. (Accessed March 3, 2023 at https://brookfieldinstitute.ca/wp-content/uploads/BrookfieldInstitute-PIP-Landscape-1.pdf.)

[10] Bernardo Gois FN, Lima A, Santos K et al. Predictive models to the COVID-19. Data Science for COVID-19 2021, 1–24. (Accessed March 3, 2023 at https://doi.org/10.1016/B978-0-12-824536-1.00023-X.)

[11] Mervosh S, Lu D, Swales V. See which states and cities have told residents to stay at home. The New York Times, 2020. (Accessed March 3, 2023 at https://www.nytimes.com/interactive/2020/us/coronavirus-stay-at-home-order.html.)

[12] Puffert D. Path dependence. EH.Net Encyclopedia, 2008. (Accessed March 3, 2023 at http://eh.net/encyclopedia/path-dependence/.)

[13] Cannabis policy simulation lab. Cannabis Public Policy Consulting. (Accessed March 3, 2023 at https://www.cannabispublicpolicyconsulting.com/cannabis-policy-simulation-lab/.)

[14] Sofis M, Slade M. Future adult use cannabis demand & predictive modeling: A behavioral economic study. Cannabis Public Policy Consulting, 2023. (Accessed March 3, 2023 at https://mgaleg.maryland.gov/meeting_material/2023/scr%20-%2013317423451784725 5%20-%20Market%20Study%20Report_01052023.pdf.)

[15] Schaneman B. 10 marijuana industry trends to watch for in 2023. MJBiz Daily, 2023. (Accessed March 3, 2023 at https://mjbizdaily.com/10-cannabis-industry-trends-to-watch-for-in-2023/.)

[16] A national weed glut is causing prices to plummet and imperiling businesses. Politico, 2022. (Accessed March 3, 2023 at https://www.politico.com/news/2022/12/25/weed-prices-business-michigan-00075485.)

[17] 2019 Recreational marijuana supply and demand legislative report. Portland, Oregon, Oregon Liquor Control Commission, 2019. (Accessed March 3, 2023 at https://www.oregon.gov/olcc/marijuana/Documents/Bulletins/2019%20Supply%20and%20Demand%20Legislative%20Report%20FINAL%20for%20Publication(PDFA).pdf.)

[18] Oregon Liquor Control Commission listening tour. Oregon Liquor Control Commission, 2019. (Accessed March 3, 2023 at https://www.oregon.gov/olcc/marijuana/Documents/Listening_Sessions/Marijuana_Listening_Sessions_Handouts.pdf.)

[19] Permanent administrative order OLCC 201-2022. Salem, Oregon, Office of the Secretary of State, 2022. (Accessed March 3, 2023 at https://www.oregon.gov/olcc/marijuana/Documents/Rules/MJ-License-Moratorium.pdf.)
[20] 2023 Recreational marijuana supply and demand legislative report. Oregon Liquor and Cannabis Commission, 2023. (Accessed March 3, 2023 at https://www.oregon.gov/olcc/Docs/reports/2023-Supply-and-Demand-Report.pdf.)
[21] Cannabis MSO Trulieve ceasing operations in Massachusetts, MJBizzDaily, 2023. (Accessed June 13, 2023 at https://mjbizdaily.com/cannabis-mso-trulieve-ceasing-operations-in-massachusetts/.)
[22] Cannabis Regulators Association (cann-ra.org)

Jahan P. Marcu, Jason B. Schechter

The role of FDA in defining *Cannabis* and hemp: A brief perspective on the future of *Cannabis* regulation in the United States

The names bestowed on plants by ancient Greeks and Romans I commend, but I shudder at the sight of most of those given by modern authorities: for these are for the most part a mere chaos of confusion, whose mother is barbarity, whose father dogmatism, and whose nurse prejudice. – Carl Linnaeus, 1737

You may call me a politician, but I am a scientist who appreciates the political process that allows me to ask tough questions and get straight non-political answers. – C.E. Turner (Marijuana Symposium, 1984)

Introduction

Scarcely anything in drug policy is more misunderstood than the scheduling process for drugs, save perhaps the role of the Food and Drug Administration (FDA) in it. This book chapter strives to explore the past and current roles of the FDA regarding *Cannabis* – including medical *Cannabis*, hemp-derivatives and, more recently cannabidiol (CBD)-dominant products, as well as possible regulatory futures for *Cannabis* once it is officially outside of schedule I (see Figure 1 and Figure 2). Ultimately, it seems unlikely that the FDA would not continue to have a key, and lasting, role in both *Cannabis* and hemp regulation. The FDA's probable role likely will depend on both prevailing funding appropriations and where *Cannabis* – in all its possible forms and derivatives – lands within the regulatory terrain. Further determinations about which products are drugs versus which are considered foods, cosmetics, herbal supplements, textiles, or something else entirely, will ultimately include some manner of FDA involvement.

The FDA plays a vital societal role in safeguarding public health and ensuring that consumer products are regulated "to meet high standards of safety and efficacy" (see www.FDA.gov). The FDA's primary role, therefore, is to protect and promote public health by ensuring the safety, efficacy, and security of human and veterinary drugs, vaccines, medical devices, food supply, cosmetics, and other products that fall

Acknowledgments: The authors extend their deep appreciation to Shane Pennington, Esq., Howard Sklamberg, Esq., Anthony Waclawski, Ph.D., and Teresa Simon – for their invaluable guidance and thoughtful recommendations that enriched the quality of this article.

Jahan P. Marcu, Marcu Enterprises
Jason B. Schechter, Cortical Systematics LLC

https://doi.org/10.1515/9783110677508-002

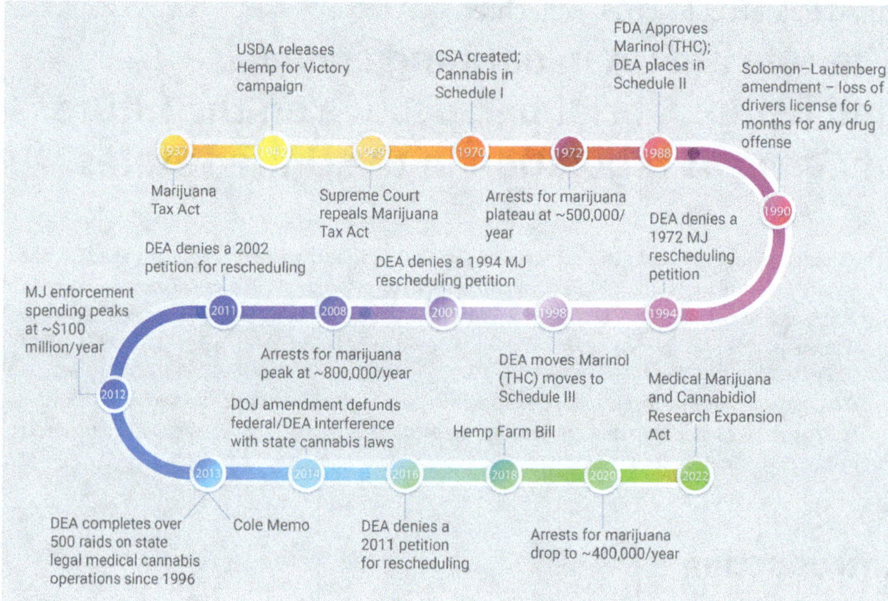

Figure 1: Federal Timeline.

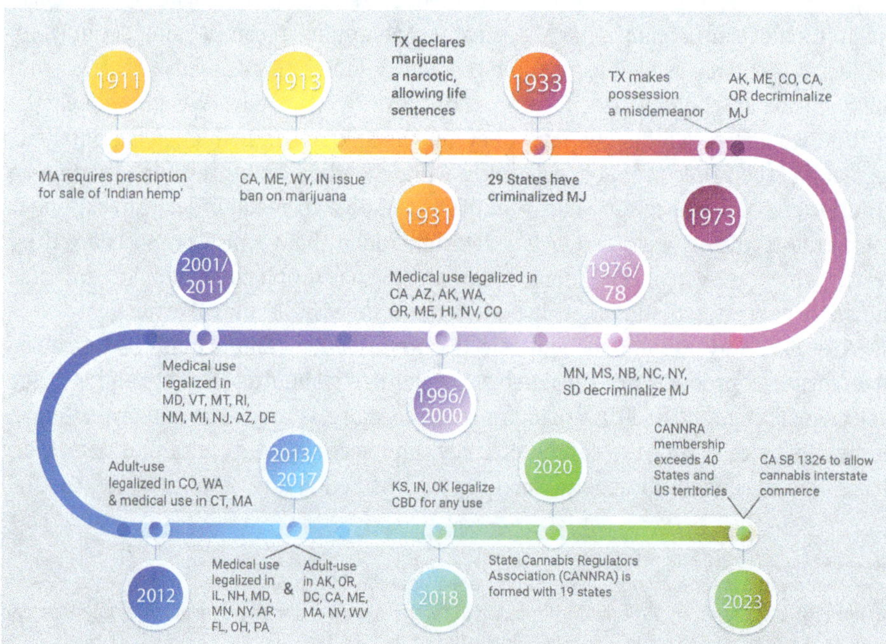

Figure 2: State Timeline.

under its jurisdiction. As such, they are responsible for regulating the development, testing, approval, manufacturing, labeling, and distribution of products for consumer use and consumption. In pursuit of this monumental goal of regulating over 1 trillion dollars in products each year, the FDA processes set standards and guidelines for product safety, conducts and oversee inspections and audits, enforces regulations, and takes action against companies that violate its rules. By conducting research and through collaborations with other regulatory agencies and international counterparts, the FDA works to inform and educate the public about health-related topics and emerging issues.

The FDA, as it exists today, evolved dramatically from the Pure Food and Drug Act in 1906. At that time, only *one* single person had an oversight role in what was just a single department in a small office. In 1938, the Federal Food, Drug, and Cosmetics Act (FDCA) furthered and expanded the reach and powers of the FDA. Over the next decade, for various reasons, the FDA successfully removed over 100 products from the US Pharmacopoeia, including *Cannabis* preparations in 1942. The FDCA also gave FDA the authority to establish Good Manufacturing Practices (GMP) for ensuring product quality. Obtaining and maintaining GMP status is required for many products such as pharmaceutical products. Subsequent legislation further expanded the role of the FDA, including the Kefauver-Harris Amendments of 1962 and the Food Safety Modernization Act of 2011, to help further ensure appropriate safety for a plethora of food and drug products under its purview.

While the lofty goals of the FDA have been historically both sprawling and well-meaning, the lens of time does suggest that the FDA's agenda has been influenced by both funding and politically-motivated decisions, irrespective of the prevailing modern science. Appropriations determined by Congress – for policymaking or regulatory enforcement activities for various substances – have been instrumental in crafting the enforcement environment for drugs of common abuse.

A long time ago, under a US presidential administration that seems so very far, far away . . . the Controlled Substance Act (CSA) was created in the 1970s by the Nixon administration. President Nixon perceived "marijuana" – an existing, demonizing slang term for *Cannabis* preparations – as "one of the greatest threats to public health", and arguably to his political power [1]. Basically, President Nixon thought people who smoked "marijuana" would not vote for him – generalizing people identifying as hippies, pacifists and political leftists, and also racial minorities, and immigrants – all groups he inappropriately, and purposely, associated with both elevated crime and drug use. A report in Harper's Magazine features this stirring quote from Nixon's domestic policy advisor:

> *The Nixon campaign in 1968, and the Nixon White House after that, had two enemies: the antiwar left and black people. You understand what I'm saying? We knew we couldn't make it illegal to be either against the war or black, but by getting the public to associate the hippies with marijuana and blacks with heroin, and then criminalizing both heavily, we could disrupt those communities. We could arrest their leaders, raid their homes, break up their meetings, and vilify them night after night on the evening news. Did we know we were lying about the drugs? Of course we did.*

President Nixon rejected prevailing evidence, both scientific and social research and reports on these subjects that did not fit his views or personal political needs. Indeed, Nixon formed the National Commission on Marijuana and Drug Abuse (aka "the Shafer Commission"), ostensibly for the purposes of collecting modern scientific findings on *Cannabis,* up to that moment. Telegraphing his wishes to the commission, Nixon can be heard in a raspy and authoritative tone on White House tape recordings:

> *I want a goddamn strong statement on marijuana . . . I mean one on marijuana that just tears the ass out of them.* [2]

Much to Nixon's disappointment, however, the extensive investigations summarized by the Shafer Commission's 1972 report – the document entitled "Marihuana, a Signal of Misunderstanding" – clearly stated that *Cannabis* should likely be decriminalized, and that prosecuting *Cannabis* users was a significant and unneeded distraction from the fight against heroin – what we would identify decades later, and still today, as a true and deadly [opioid] epidemic [3]. Nixon was infuriated, later stating:

> *You know, it's a funny thing, every one of the bastards that are out for legalizing marijuana is Jewish.* [4]

To be clear, both then and now, Federal "marijuana" conviction results in the loss of personal rights, including losing the constitutional right to vote in the United States – euphemistically termed "voter disenfranchisement." The resulting disproportionate socio-economic impact of Nixon's drug policies on communities of color has today been laid bare in our society, while only modest attempts have been made to correct such injustices with social equity programs.[1] By calling *Cannabis* "marijuana" – in an attempt to "Mexicanize" the plant in the eyes of an otherwise ignorant public – by investing in a global fear campaign, and by placing cannabinoids in Schedule I status (i.e., "no medicinal use"), Nixon's anti-*Cannabis* maneuvering was an expedient ploy for maintaining political power. The most efficient way to promulgate the political power through the Federal prohibition of "marijuana" Nixon sought was to appoint *law enforcement* to be in charge of both assessing *and* responding to "marijuana's" so-called "threat to society". Therein, too, lies the origin of the Nixon's Drug Enforcement Agency (aka DEA) in 1973, and the source of its inherent power – to both *make* drug scheduling decisions and *enforce* them. Each of these powers originated from Nixon's desire for expediency and motives that were coherently, unapologetically political.

Hopefully, this somewhat clarifies *why* a nonscientific organization, unelected by the public, with no experience in either generating or assessing drug abuse research (or health data in general), would somehow be trusted to determine so-called public health threats for a nation of 331 million people . . . while *also* possessing unprece-

1 See Michelle Alexander's book The New Jim Crow for an account of racial injustices and important perspectives related to drug laws.

dented authority to mitigate these threats (real or imagined) with true paramilitary force (e.g., raiding medical *Cannabis* operations and patient cooperatives of infirm cancer survivors). Today the role of reviewing data and information about abuse potential of product to be scheduled is given to the U.S. Department of Health and Human Services (HHS) by Congress and this role is delegated to the FDA. Under USC 811(b), the scientific and medical views of the HHS and the FDA can be binding on the DEA in the scheduling process – Compromise between the HHS, FDA, and DEA is the ultimate decision maker. Onwards to the *how* . . .

It is often easier to talk about drug substances using scheduling categories, if not only because there are *so* many drugs that exist in our pharmacopeia, with more discovered and formulated and coming to our attention every day. Further, the regulatory status of individual drugs changes over time and may be different even across jurisdictions. In general, the more restrictive the scheduling status of a drug, the more restrictive or prohibited is *access* to that substance for research, manufacturing, and sale at the Federal level. Further, the mobility of drugs *between* various scheduling categories is further hampered by agencies that often have little or no real expertise regarding the manifold criteria for (re-)scheduling, such as assessing the *quality* of scientific information or the effects of suppression of purified research materials for investigation.

Presently there are five (5) scheduling categories, created for the CSA. The categories of "I" through "V" are in descending order of restrictiveness. In the simplest of terms, "Schedule I" criteria are substances that have "no currently accepted medical value and a high potential for abuse."(Figure 3 shows a breakdown of the scheduling status.) Schedule I drugs are also considered to be *so* dangerous that possessing that status prohibits any *"use of the drug or other substance under medical supervision."* "Schedule II" drugs are those compounds that have a "high potential for abuse" or "severe psychological or physical dependence", though there may also be some accepted medical use with deliberate limitations. "Schedule III" further slackens the abuse and safety criteria, and these compounds are substances that "have less abuse potential" than Schedule I or II substances, and only a *"moderate or low physical dependence or high psychological dependence."* "Schedule IV" has still lower risks of abuse and other concomitant factors. Finally, "Schedule V" compounds ostensibly possess the lowest concern of drug safety criteria for scheduled drugs. Drug compounds with lower potential for abuse than Schedule IV typically consist of a low potency formulation of a drug or drug type found in other scheduling categories (some advocates are lobbying for creation of a further, sixth scheduling category for certain products, discussed below in the section Pathways for Federally Legal *Cannabis*). Terms such as "abuse potential" and "dependence" do not hinge on the actual danger of the drugs, which is perhaps a confusing aspect of scheduling – that drugs that seem more dangerous are in less restrictive categories than those that seem less dangerous. Schedule II or III drugs can be just as dangerous or even more dangerous than Schedule I drugs, the only difference being a currently accepted medical use.

Drug scheduling determinations are made by narrowly selecting data, with analysis about abuse potential and "safety" that is neither compared to, nor in the societal context of, the other common drugs of abuse in their respective scheduling categories. More succinctly, scheduling can be described as an absolute versus relative risk approach – and, as such, a serious flaw in our drug regulatory system may be that we seldom, if ever, see agencies conduct either an inter-, or intra-, scheduled-drug comparison to inform wide-ranging policy changes. Without such a siloed and subjective scheduling process, our government would arguably have to laboriously compare *individual drugs* to either *other* substances within each category or disparate substances within the same category. For example, instead of attempting to fit the definition of a scheduling status to the drug, the DEA would have to do something akin to comparing *Cannabis* (in Schedule I) to fentanyl (in Schedule II) or ketamine (in Schedule III) to make scheduling determinations. It is hard to believe that the US government does not formally examine products such as THC (the most active molecular component of *Cannabis*) and ask if it is *either* more dangerous *or* more addictive than other approved drugs (such as fentanyl or ketamine) before making such a wide-ranging scheduling decision. Perhaps this political, intellectual laziness is why THC appears in *three* different scheduling categories but can also be found as an unscheduled substance (see Figure 3).

After the DEA places a drug in a scheduling category, it may not be the end-all placement for that substance (see Figure 1 for examples of how cannabinoids have moved around in scheduling status). The CSA specifically authorizes the Attorney General to add or remove items from the scheduling, including moving items from one schedule to another, with the DEA as the implementing agency [5]. Any changes must be guided by the CSA's process, which includes requesting both a scientific and medical evaluation, and requires a recommendation from the HHS. Responsibility for the evaluation and recommendation is delegated, in turn, to the FDA. Federal agencies, such as the FDA, are continuously exploring and evaluating various approaches to regulating *Cannabis*, *Cannabis*-derived and hemp-derived products, and compounds, to little avail.

Before proceeding with any actions to loosen control or remove *Cannabis* from the schedules established in the CSA, the DEA is allegedly required to collect necessary data, seek a scientific and medical evaluation from the HHS and FDA, and request a scheduling recommendation from the HHS. Again, as this chapter was being written there has been an HHS recommendation to move *Cannabis* to Schedule III, which has landed in front of the DEA. And the DEA *may* now be taking strides – though the DEA, historically, can be seen to have implemented a somewhat geological-time, slow-walk approach to review anything regarding *Cannabis.* Indeed, the DEA took *5 years* to respond to the last petition to reschedule *Cannabis*, well over *20 years* for the previous petition, and over *4 years* of delays before reviewing a single license application to grow research *Cannabis* for the federal government.[6–8] Basically, the HHS and FDA confer on a schedule recommendation, and then the DEA reviews the petition or rec-

Drug	Schedule Status	Year Scheduled	Interventions	FDA Approval
Marijuana/*Cannabis* and extracts	I	1970	Several failed attempts to have a hearing on rescheduling	No
Tetrahydrocannabinols (THC), derivatives, and isomers (i.e., D8-THC)	I	1970		No
THC (Marijuana, *Cannabis*)	I	1970		No
CBD, derivatives, and isomers	I	1970	Any CBD-type products with a concentration of > 0.3% THC are schedule I drugs	No
Syndros (THC)	II	2016		Yes
Cesamet (THC isomer)	II	1985	1989 Discontinued by Lilly for commercial reasons	Yes
Marinol (THC)	III	1985	1998 Moved from Schedule II to Schedule III	Yes
Epidiolex (CBD)	Not Scheduled	2018	2020 Moved from Schedule V to completely descheduled	Yes
THC (hemp)	Not Scheduled	n/a	2018 Farm Bill defines hemp as *Cannabis* with no more than 0.3% THC on dry weight basis	No
CBD (hemp)	Not Scheduled	n/a	Farm Bill (see above)	No

Figure 3: Scheduling Status of THC, CBD, Hemp, *Cannabis*, and Marijuana.

ommendation, and then the DEA implements a decision. It is easy to understand why *Cannabis* rescheduling decisions are slow moving – indeed, we all have things that we do not want to do for deeply enshrined, financially-inconvenient political reasons, that then become a low priority. Rinse. Repeat.

At the time of writing this, *Cannabis* (or, tragically, somehow *still* "marijuana") is *still* a "Schedule I" drug, the most federally restrictive designation possible – irrespective of the *thousands* of known uses for cannabinoid compounds and admixtures thereof and nary a single human mortality. It has been several months since President Biden's announcement on October 6, 2022 to "expedite" a scheduling review of *Cannabis* [9]. President Biden's directive instigated departmental activity amongst the HHS and the U.S. Department of Justice (DOJ), the U.S. Drug Enforcement Administration (DEA), and the U.S. Food and Drug Administration (FDA). Although HHS issued a recommendation for *Cannabis* to be placed in Schedule III – a comparatively less restrictive category[10] – it has apparently yet to be seriously considered.

What resources can the FDA use?

The FDA has several mechanisms at its disposal to help the organization discharge its responsibilities of regulation and enforcement. These include setting standards and requirements for products (aka "rules and regulations"), scheduling and executing facility inspections, facilitating product reviews and approvals, conducting labelling reviews, and by testing and sampling products. The FDA can detain and/or refuse imports, implement recalls and seizures, orchestrate public education programs, publish guidance documents, and can issue warning letters to companies in violation of Federal regulations.

It could be argued, that with all of these mechanisms and responsibilities under its purview, perhaps the FDA should not be the government entity to deal with as large and wide-ranging a subject matter as *Cannabis* products and medicines. As such, a remarkable catalyst for fervent discourse about *Cannabis* is to ask if FDA enforcement might be increasing or decreasing for *Cannabis* and hemp. While data indicates that total formal court enforcement of *Cannabis*-related issues is decreasing across all things the FDA regulates, this form of enforcement has been replaced with different forms of compliance programs.

The FDA can be seen to use a fair amount of informal enforcement techniques, which include premarket approval programs that invite extra scrutiny, regulations, preambles, guidance documents, and warning letters. By far the most popular of these informal enforcement options is use of the "warning letter" – a technique used regularly for hemp and CBD products (see Table 1). The purpose of these programs is to mitigate any ambiguity that formerly would have led to legal interventions and court enforcement. The FDA's power derives, in part, from public perception and trust in this authority.

Another effective, informal compliance activity is reputation-driven, tactical use of the media by the FDA. Effective, if one-sided, press releases and other public statements greatly influence both the public and industry sectors, without the opportunity for any public participation. Several times throughout its short history, the FDA has used publicity to impact an ingredient's reputation: think cranberries, or nonfat dry milk, or phenylpropanolamine . . . ephedra, opioids, e-cigarette/vaping products, artificial sweeteners, genetically modified organisms, trans fats, etc. Direct marketing of FDA data is an effective mechanism of propaganda, even if irrespective of scientific rigor. Albeit, the effectiveness of such media tactics can be negligible if the targeted groups simply do not care about their reputation in the absence of formal court actions. Indeed, money remains an excellent motivator in continuing production of known dangers to public health.

Table 1: Enforcement Actions Against CBD, Hemp, and *Cannabis* Products Enforcement Actions.

Date	Title	Document type
July 2023	FDA, FTC warn six Companies for illegally selling copycat food products containing Delta-8 THC	FDA news release
January 2023	FDA concludes that existing regulatory frameworks for foods and supplements are not appropriate for cannabidiol, will work with Congress on a new way forward	FDA statement
January 2023	FDA issues response to three citizen petitions related to CBD and dietary supplements	CFSAN constituent update
November 2022	FDA warns companies for illegally selling food and beverage products that contain CBD	CFSAN constituent update
May 2022	FDA warns four companies for illegally selling CBD products intended for use in food-producing animals	CVM update
May 2022	FDA issues warning letters to companies illegally selling CBD and Delta-8 THC Products	FDA news release
March 2021	FDA warns companies illegally selling over-the-counter CBD products for pain relief	FDA news release
December 2020	FDA warns companies illegally selling CBD products	FDA news release
July 2020	FDA approves new indication for drug containing an active ingredient derived from *Cannabis* to treat seizures in rare genetic disease	FDA news release
April 2020	FDA warns companies illegally selling CBD products to treat medical conditions, opioid addiction	FDA news release
March 2020	FDA advances work related to cannabidiol products with focus on protecting public health, providing market clarity	FDA statement
November 2019	FDA warns 15 companies for illegally selling various products containing cannabidiol as agency details safety concerns	FDA news release
October 2019	FDA, FTC warn company marketing unapproved cannabidiol products with unsubstantiated claims to treat teething and ear pain in infants, autism, ADHD, Parkinson's, and Alzheimer's disease	FDA news release
July 2019	FDA warns company marketing unapproved cannabidiol products with unsubstantiated claims to treat cancer, Alzheimer's disease, opioid withdrawal, pain, and pet anxiety	FDA news release
July 2019	FDA is committed to sound, science-based policy on CBD	FDA voices
April 2019	Statement on new steps to advance agency's continued evaluation of potential regulatory pathways for *Cannabis*-containing and *Cannabis*-derived products	FDA statement

Table 1 (continued)

Date	Title	Document type
December 2018	Statement on signing of the Agriculture Improvement Act and the agency's regulation of products containing *Cannabis* and *Cannabis*-derived compounds	FDA statement
June 2018	Statement on the importance of conducting proper research to prove safe and effective medical uses for the active chemicals in marijuana and its components	FDA statement
June 2018	FDA approves first drug comprised of an active ingredient derived from marijuana to treat rare, severe forms of epilepsy	FDA news release
November 2017	FDA warns companies marketing unproven products, derived from marijuana, that claim to treat or cure cancer	FDA news release

Table adapted from FDA website: **FDA Regulation of *Cannabis* and *Cannabis*-Derived Products, Including Cannabidiol (CBD)** https://www.fda.gov/news-events/public-health-focus/fda-regulation-cannabis-and-cannabis-derived-products-including-cannabidiol-cbd

Past and present – what the FDA is currently doing with *Cannabis* and hemp

The FDA has stated that *Cannabis* products, whether classified as hemp or not, are subject to the same authorities and conditions as FDA-regulated products containing any other substance [11]. According to the FDA, standard drug approval processes represent the best way to ensure the safety and efficacy of any drugs or substances derived from *Cannabis* or hemp [12]. A former FDA Commissioner laid out some compelling reasons why FDA should continue to regulate not only hemp products, but any product containing substances classified as a drug by the FDA – which includes THC and cannabidiol (CBD) – on which medical claims are made, regardless of scheduling status [13].

While *Cannabis* remains in Schedule I, it can be argued that one of the primary roles of the FDA is to support both scientific research and drug development of the cannabinoids. In 2017, it was stated by a representative from the agency that the "*FDA has a clear role with multiple activities around the regulation of marijuana.*" [14] This ongoing work includes providing scientific advice on the risks of *Cannabis* and its constituents (i.e., 8-Factor analysis). This role also, however, includes supporting rigorous scientific research into any therapeutic value of *Cannabis* and its constituent cannabinoids. In perhaps its most visible role, the FDA is publicly seen taking appropriate actions related to the marketing of products containing both *Cannabis* and its constituent cannabinoids (see Table 1 for list of recent actions). These are epitomized as FDA warning letters to CBD companies (discussed further in the upcoming sections).

Today's FDA appears committed to streamlining *Cannabis* product oversight, to speed the availability of "*new drugs from marijuana for the American public.*" It is interesting that the FDA intends to do all this, and yet there is no dedicated, voluntary, national system to collect *data* on adverse events or consumer experiences with *Cannabis* products. The current FDA Adverse Event Reporting System (FAERS) is crippled with limitations and use issues[2] – but for the time being it may be the best mechanism there is to detect safety signals related to consumer products [15, 16].

The FDA seems to have consistently treated *Cannabis*-based products as "new drugs"[3] under the FDCA, which means that a *Cannabis* product cannot be transported for commerce between states without either an approved new drug application (NDA – which permits commercial marketing) or an investigational new drug applications (IND) approval for clinical testing. The IND application, in particular, is an application for an *exemption* from the law, which prohibits interstate commerce related to an unapproved drug. Interstate commerce is a foundational concept for the FDA, because it grants the agency the authority to regulate a wide array of products, including the movement and distribution of goods, food, drugs, medical devices, dietary supplements, cosmetics, and other regulated products across State lines (within the United States). At a fundamental and basic level, interstate commerce is a cornerstone of a strong Federal government's power (discussed at length in the Federalist Papers[4]).

Perhaps the most infamous INDs[5] were created in the 1970s, as the agency permitted INDs for the use of *Cannabis* to treat a variety of ailments, including refractory glaucoma and anorexia associated with AIDS [17]. Although the IND regulatory mechanism normally governs clinical trials, the INDs permitted here were for so-called "compassionate use" programs. Unlike clinical trials, these programs *were not intended* to generate data for purposes of new drug approval. Rather, they were used, simply to permit treatment of individual (and rare) patient's disease or condition with a Schedule I drug. The *Cannabis* IND program was established initially as a

---

2 The FAERS database is criticized for how it categorizes and spells certain drugs which must be spelled exactly, or they do not generate a search result. For example, psilocybin is spelled as "psilocybine" in FAERS. A sign of hope is that the cannabinoid terminology, which is ill-defined and yet seems to contain many options for *Cannabis*-derived products is overrun with potentially redundant terms. This makes it difficult to cross-reference cases reported to FAERS.

3 New drug substance means any substance that when used in the manufacture, processing, or packing of a drug causes that drug to be a new drug, but does not include intermediates used in the synthesis of such substance.

4 The Federalist Papers played a significant role in the ratification of the U.S. Constitution by helping to persuade the public of the advantages of a strong federal government. The regulation of interstate commerce is a key component of the U.S. Constitution. The Commerce Clause, found in Article I, Section 8, Clause 3 of the Constitution, grants Congress the authority to regulate commerce among the states.

5 IND contents are considered confidential.

settlement to a civil lawsuit involving Robert Randall [18]. A court ruled his "use of marijuana" was a "medical necessity," dismissing charges against him. This was followed, as one would imagine, by a fair amount of "bureaucratic wrangling" until the Federal government agreed to supply him with *Cannabis* (in pre-rolled cigarette form) to smoke for his glaucoma.[6] *So is Cannabis medicinal or not?* The solution to this dilemma was for the Federal government to create this Compassionate Use Program, launched in 1976 [19] as part of an existing program to provide seriously ill patients with *Cannabis* and promising medicines prior to their approval by the FDA.[7]

While over a *hundred* people sought out this government service, very few were approved – the great majority of applicants were denied access to *Cannabis* medicine. (See Kuromiya v. United States for more detail about the closure of this program [20].) While these early "treatment INDs" are *technically* no longer active, the survivors of the initial program were grandfathered and, as such, *continue* to receive a pound of pre-rolled *Cannabis* cigarettes – approx. 300 pre-rolls each month – until their death.[8] Via this mechanism, the FDA *has effectively already permitted* dozens of INDs for true clinical trials of *Cannabis* and *Cannabis*-based products. Through the lens of this precedent, it is entirely likely the FDA could view most *Cannabis* products as "drugs" and initiate them for placement within the CSA (i.e., between Schedule II–V), which would then ultimately be determined by DEA.

The FDA has a standardized report template to evaluate individual drugs for placement as controlled substances. This standardized approach in manifested in the aforementioned "8-Factor analysis" (outlined in 21 U.S.C. 811). The 8-Factor analysis is a written report, which ultimately includes a scheduling recommendation. Shortly after the last *Cannabis* rescheduling effort, the National Academies of Sciences, Engineering, and Medicine (NASEM) released a report which seemed to be in stark con-

---

6 From Russo 2002 "The material he received from his study physician was cultivated in a 5-acre plot at the University of Mississippi, mostly from seeds of Mexican origin, and was rolled and packaged at the Research Triangle Institute in North Carolina under the supervision of the National Institute on Drug Abuse (NIDA). . .Through the Freedom of Information Act (FOIA), he discovered in 1978 that the government's cost of *Cannabis* cultivation and production was 90 cents per ounce (28 g), with 2/3 of this cost attributable to security measures. Thus, the actual cost of production approximated 1 cent per gram (US $0.01/g)."

7 From Mack & Joy 2001 "While the Compassionate Use program for marijuana smokers now exists solely as a historical artifact, the medical necessity defense remains viable for some patients who treat their symptoms with marijuana. This defense originates from the common law principle that illegal actions are excusable or justified if they are taken to avoid even greater harms. Courts considering such cases must balance the interest of the individual patient against the government's interest in upholding the law."

8 The IND program allowed access for individuals to approximately 300 grams of a Schedule I drug (a drug by definition that has no medical use) per month because it was a medical necessity.

trast to the FDA's stance. The NASEM report on *Cannabis* report considered over *10,000* published studies, as compared the last 8-Factor analysis, which contained only a few hundred studies. The NASEM report, unsurprisingly, found "substantial evidence" supporting the use of *Cannabis* for multiple health conditions – counter to the findings of the FDA, HHS, and DEA. When the DEA published their final ruling with input from HHS and FDA, it included only *three* (3) clinical studies to inform its decision to maintain the status quo of *Cannabis* in Schedule I. Additionally, they continue to call it "marijuana."

To date, the FDA has not yet approved a single marketing application for a state legal *Cannabis* product for the treatment of any disease or condition. To clarify, **no product currently sold through a State licensed dispensary has yet been evaluated or approved by the FDA**. The agency has, however, approved one (1) *Cannabis*-derived drug product: Epidiolex™ (cannabidiol), and three (3) synthetic *Cannabis*-related drug products: Marinol™ (dronabinol; Schedule III), Syndros™ (dronabinol; Schedule II), and Cesamet™ (nabilone; Schedule II). These approved drug products are only available with a prescription from a licensed healthcare provider. Importantly, the FDA has not approved any other *Cannabis*, *Cannabis*-derived, or cannabidiol (CBD) products currently available on the market. The FDA has approved four (4) generic dronabinol products presented in oral capsules at the same strength as Marinol™ (see Table 2).

Table 2: Approved THC Products.

Drug name	Company name	Application number	Dose	Form	Marketing status	Active ingredient
Marinol	Alkem Labs Ltd.	018651	2.5, 5, 10 mg	Capsule; oral	Prescription	Dronabinol (tetrahydrocannabinol)
Dronabinol	Ascent Pharms Inc.	207421	2.5, 5, 10 mg	Capsule; oral	Prescription	Dronabinol (tetrahydrocannabinol)
Dronabinol	SVC Pharma	078292	2.5, 5, 10 mg	Capsule; oral	Prescription	Dronabinol (tetrahydrocannabinol)
Dronabinol	Akorn	079217	2.5, 5, 10 mg	Capsule; oral	Prescription	Dronabinol (tetrahydrocannabinol)
Dronabinol	Insys	078501	2.5, 5, 10 mg	Capsule; oral	Discontinued	Dronabinol (tetrahydrocannabinol)
Dronabinol	Lannett Co Inc.	201463	2.5, 5, 10 mg	Capsule; oral	Discontinued	Dronabinol (tetrahydrocannabinol)
Syndros	Benuvia Operations	205525	5 mg/ml	Solution; oral	Prescription	Dronabinol (tetrahydrocannabinol)

There are indeed some limitations to the FDA's potential reach into *Cannabis* regulation, though, as FDCA powers enacted under the Commerce Clause permits the FDA to regulate only products that travel or have traveled in interstate commerce. Specific *Cannabis* products may fall out of the FDA's purview and therefore be the responsibility of the individual States. For instance, whole *Cannabis* plant/flower products (produced, grown, and sold in the same State) not marketed with a health-related claim (e.g., vs. pain, anxiety, sleep) would fall out of the FDA's field of view. However, FDA enforcement action *has* been taken against both hemp and *Cannabis* businesses, emboldened by State legalization, with diverse product categories and overt marketing and/or sales activities in interstate commerce. The issue of interstate commerce may be addressed as States begin enacting their own legislation to allow *Cannabis* to be transported from one jurisdiction to another, such as California's SB1326 [21]. It is however doubtful that States would *truly* have full power to carve out specific interstate commerce rules, as regulating interstate commerce is the mainstay of power for the Federal government. Indeed, there is a good argument that regulation of interstate commerce is the true responsibility of the Federal government as one of the predominant reasons for its existence. Regardless, simply rescheduling or descheduling *Cannabis* and/or cannabinoid compounds would not inherently change either the FDA's resources or authority to take compliance action where appropriate or would it give the agency additional enforcement resources. Put differently, no appropriations or additional authorities would be granted to FDA simply by virtue of either moving *Cannabis* out of Schedule I or descheduling *Cannabis* altogether (which could arguably include *all* its products and derivatives).

What does the FDA currently do with *Cannabis*, cannabidiol, and hemp?

FDA has compliance staff in the Office of Regulatory Affairs, the Center for Food Safety and Applied Nutrition, and the Center for Drug Evaluation and Research (CDER). CDER, specifically, is engaged in surveillance, compliance, and enforcement activities, including those meant to minimize consumer exposure to unsafe, unapproved drugs, misbranded and adulterated products, etc. Such offices have initiated compliance actions (e.g., issuing warning letters) against both individuals and companies who are marketing CBD (and Delta-8 THC) products in violation of the FD&C Act. FDA can and does use its resources to take enforcement action, typically against individuals or entities who commit encroachments on the FD&C Act and/or against violative products.

Following the interest in hemp-derived CBD that the 2018 Farm Bill generated, the FDA increased its focus on CBD. In response, the FDA formed a high-level workgroup dedicated to coordinating our approach to CBD policy-making. The FDA stance on

CBD and Epidiolex™ highlights a certain degree of comparative leniency when it comes to products that seemingly violated the FD&C Act – even after granting approval to a product containing the *identical* active ingredient (many will argue that hemp extracts and hemp oil products are distinct from Epidiolex™). The FDA's enforcement efforts regarding CBD, to date, have primarily focused on products that make significant health claims, especially those that have potential negative impacts on public health or are apparently marketed towards children. The FDA has explicitly stated its intention to continuously monitor the market and take appropriate measures to safeguard public health from the illicit sale of hemp, *Cannabis*, *Cannabis*-derived, and hemp-derived products that might jeopardize consumer well-being – particularly those marketed for unapproved therapeutic purposes. Such a widely accepted and reasonable approach is very likely to persist even after any rescheduling or descheduling event.

The FDA has largely enforced (or taken compliance actions) against hemp companies for label and marketing claim violations. Past warning letters have cited companies illegally selling CBD products that illegitimately claimed to "prevent, diagnose, treat, or cure serious diseases," such as cancer – noting that these are ostensibly "drug" claims (see Table 1 for examples). Some of these products were also, according to FDA, in violation of the FD&C Act, as they were marketed as "dietary supplements" or because they involved the simple addition of CBD to food. FDA has generally not taken compliance action against CBD products that do not make drug claims and do not target children.

On January 26, 2023, the FDA concluded that hemp, CBD, and derivatives thereof, cannot be regulated as a dietary supplement. The FDA declined to issue any regulations to allow CBD as dietary supplement and suggested a new agency be formed, with the sole focus of hemp-CBD regulatory compliance and enforcement [22, 23].

Pathways for federally legal *Cannabis* following re(or, maybe, de-)scheduling

A more robust regulatory system would do well to address the same priorities as the Department of Justice – as proper enforcement should surely include the things most of *Cannabis'* staunchest critics are worried about, such as prohibiting access to minors, preventing illicit trade from funding criminal enterprises, human trafficking of illegal workforces, black market violence, cash-only business susceptibility, etc. For individual *Cannabis* products and derivatives, all regulatory pathways remain open – paralleling the pathways that led us to the approval of Epidiolex™ and Marinol™. Whether or not the true source material, the *Cannabis* plant itself, ultimately ends up in a scheduled category and is entirely *separate* from its status as an approved product. Indeed, *Canna-*

bis contains a *vast* array of active molecules, as well as various terpenes and flavonoid constituents, each of which is likely prone to regulatory processes.

These complexities mean that regulatory agencies face significant challenges regarding *Cannabis*. Below we briefly discuss some possible regulatory outcomes for *Cannabis*:

Maintain existing regulatory status and structure

Some will argue that the current Federal regulatory system for *Cannabis* and hemp is working well, because it has been adequate to gain approved use for specific products. Others think it is not working and that *Cannabis* should be removed from any scheduling category and regulated entirely separately. However, a policy decision to keep all *Cannabis* federally illegal might continue to further embolden the illicit market, eventually resulting in a large portion of the *Cannabis* industry put out of business. Indeed, stress fractures in the existing system – due to economic strain, related to unfair illicit competition and banking issues related to the plant's Schedule I status (i.e., 280E tax issues) – are already evident [24]. Maintaining the *status quo* would also undoubtedly continue to perpetuate the many injustices of the "drug-war on marijuana," preventing any meaningful, or even coherent, policy that could facilitate safe, effective, and clinically validated *Cannabis* or hemp products.

Descheduling, according to legal scholars

According to legal experts and scheduling reform coalitions, "descheduling" would require a congressional amendment of the CSA to remove "marijuana" – or at least some parts of that broad statutory term – from any of the restricted Schedules of controlled substances.[8, 24–28] Descheduling is, quite literally, the *removal* of all-things *Cannabis* from any scheduling category. Descheduling would include DEA action to remove medically relevant components of *Cannabis* – including THC and CBD, as well as synthetically-produced analogues – which it has placed previously in any of the controlled substances schedules, under its *statutory* authority granted under the CSA. Full descheduling would mean that (medical) *Cannabis* would no longer be a "controlled substance" at any level and is the only result of possible reform efforts that would completely remove *Cannabis* from the DEA's purview [25]. **Even if *Cannabis* was moved to Schedule V, it could be expected that *every transaction* in the medical and adult-use market would still be illegal under federal law** and subject to penalties [24]. However, the odds seemed stacked against descheduling becoming an unhindered and boundless evolutionary phase for *Cannabis* and cannabinoids. If *Cannabis* were descheduled, we should anticipate that it would still remain, somehow, fully regulated and at times, in certain cases at a minimum, the FDA would still retain jurisdictional standing to intervene. Ergo, the FDA would likely maintain its power to act against interstate sale if *Cannabis* products were suddenly considered an unapproved drug. And descheduling *Cannabis* would likely not directly affect FDA-approved cannabinoid products, mentioned earlier.

Under descheduling, though, *Cannabis* that remains inside a U.S. State, is grown, manufactured, packaged, or otherwise used in that same jurisdiction, would likely remain outside the FDA's jurisdiction. A conceptual example might be something like the sale of a bag of raw *Cannabis* flowers by a vertically licensed operator, for personal consumption, within the same State. However, the FDA would still remain concerned about safety or claims around a product and could use any ingredient that has traveled through interstate commerce to gain jurisdictional grounds. Further, any marketing claims seen in another State, or products purchased in one State and consumed in another could constitute introduction of that product into interstate commerce (i.e., 21 USC 331). While some might suspect that a Congress willing to deschedule *Cannabis* would *also* be open to amending the FDCA – to allow the unfettered production of *Cannabis*, marketing claims without clinical data (and sales without restrictions) – a legal review suggests that "free-form expansion" of the *Cannabis* market does not necessarily follow descheduling [25].

A further underappreciated issue resulting from *Cannabis* descheduling might be a further decrease in true pharmaceutical drug development that may even hamper research into therapeutic benefits. Indeed, medical products are unlikely to be developed when the active ingredients are already readily available on a commercial market. Scheduling restricts the market for unapproved drugs, and therefore is a barrier to competition – approved drugs can travel across State boundaries *and* make marketing claims as to efficacy. It is unlikely that either investors or government funding agencies would support (i.e., pay for) any robust research and development programs that cannot lead to viable (and highly lucrative) products. Descheduling may turn *Cannabis* and hemp biotech pitches into a high-risk, low-reward venture.

Rescheduling to III–V, according to legal scholars

"Rescheduling" means that Congress amends the CSA to place "marijuana" into a less restrictive schedule status. Likewise, for full rescheduling, the DEA would also move any medically relevant natural or synthetic *Cannabis* components that it had previously placed on Schedule I further down, to either the same or lower Schedule that *Cannabis* would now occupy. Rescheduling means that *Cannabis* would *still* be a controlled substance, but it would be easier to pursue clinical trials and new drug approvals derived from the plant. Further, over-the-counter status would at least be a theoretical, if still unlikely, possibility for every product form.

Moving *Cannabis* to Schedule III, IV, or V could correct many, but not all, of the inherent wrongs regarding our present "failed approach to marijuana" [24]. A few benefits of rescheduling *Cannabis* to III-V include alleviating the glaring 280E tax penalty – which would allow regulated *Cannabis* businesses to take ordinary tax deductions, significantly decreasing the scope of criminal liability for *Cannabis*-related activity and businesses, while freeing up the courts (as prosecutors and sentencing judges would be less likely to view *Cannabis*-related activity as a high priority, criminal concern when leveling charges and/or issuing sentencing). However, none of these things would be possible if *Cannabis* were only moved to Schedule II – it would need to be Schedule III, IV, or V to have significant effect.

Moving *Cannabis* to Schedule II is arguably *not* a good idea and would be entirely unlikely to solve any issues facing the *Cannabis* industry, public health, and researchers. In point of fact, *opponents* to *Cannabis* rescheduling concerns suggest that if *Cannabis* were to be rescheduled – shifting *Cannabis* to Schedule II – it would be *unethical*, because it would do nothing to improve the situation (i.e., make research easier). To wit, a former White House drug advisor – the *sole* expert witness for the U.S. Department of Justice in the 2015 landmark Federal Court case for the descheduling of "marijuana," stated:

> *"The process for marijuana research could be streamlined by Drug Enforcement Administration oversight and expansion of marijuana production . . .* ***But moving marijuana to Schedule II "to promote research" is conceivably unethical . . .*** *The marijuana plant, like other plants, is not reasonably amenable to analysis by the current, established FDA approval process. While many of our medicines have their origin in plant- derived compounds, I am unaware of any whole plant [product] that has been approved for medical use in recent times . . . rescheduling* [Moving *Cannabis* into schedule II will not] *open the floodgates for research"* – Bertha Madras, Ph.D., Former US Deputy Drug Czar[9]

There is a common misconception that if *Cannabis* plant were to be rescheduled, it would become an "approved" product, officially under "drug" status, everywhere. It is true, that under the FDA framework, rescheduling *could* trigger a variety of regulatory requirements. However, automatically approving all *Cannabis* products and preparations as "FDA-approved" products is not what would happen. *Cannabis* (whole flower) would remain an unapproved product; a *Cannabis*-product would become an "approved product," but only if it was taken through, for example, the FDA's botanical drug approval pathway. A Schedule III designation could perhaps put more pressure on companies looking to sell products across State lines, to comply with "current good manufacturing practices" (cGMP, a requirement for drug-related interstate commerce). This would mean that failures to comply with cGMP would assume that the product is adulterated, resulting in some enforcement action [29]. Most companies in medical and adult-use *Cannabis* already follow robust quality programs, testing procedures, and various child-protective packaging requirements [12, 30, 31]. It is therefore somewhat unlikely that future FDA GMP requirements would be more severe than what is already required in many States for

9 https://www.washingtonpost.com/news/in-theory/wp/2016/04/29/5-reasons-marijuana-is-not-medicine/

medical *Cannabis* products (e.g., New York State). Several third party companies already provide GMP or GMP-like certifications for *Cannabis* and hemp companies. Further, many *Cannabis* companies have *already* achieved GMP or GMP-like status voluntarily, without it being a specific federal requirement for *Cannabis* operators.

There seems to be a lot of confusion right now about the HHS recommendation to move *Cannabis* to Schedule III. Compared to other mechanisms, though, it is actually quite simple. The recommendation can only be made by FDA's parent group, the HHS, **if** *Cannabis* has currently accepted medical use, though it lacks approval for interstate commerce. Patients have been using *Cannabis* with a doctor's recommendation *for decades* without FDA intervention, almost as if *Cannabis* was already out of Schedule I, federally. It therefore makes little to no sense that the FDA or DEA would suddenly increase spending to severely clamp down on *Cannabis* operations that are compliant with state laws after rescheduling. It is important to remember that the regulatory status of *Cannabis may* change soon (i.e., Schedule I to Schedule III) but this *in no way* means it has become an "FDA-approved product" that would require a prescription. **The *Cannabis* plant would remain an unapproved product by the FDA whether it is rescheduled or descheduled.**

In summary, moving *Cannabis* to Schedule III, IV, or V (but *not* Schedule II) would serve admirably to advance the Biden administration's goals, marking a historic step on the path to necessary reform. Rescheduling would reduce the scope of criminal liability for *Cannabis*-related activity, provide Federal tax relief from usury tax penalties associated with the enforcement of Section 280E of the Internal Revenue Code, and would provide overall support to an industry already approved by the great majority of Americans, while generating billions of dollars for the American economy, while creating hundreds of thousands of American jobs. This is clearly why *Cannabis*-policy reform proponents justly focus on rescheduling efforts, so that it would be treated like a prescription drug, rather than as a narcotic, too dangerous to use, even under medical supervision. At this moment in history, rescheduling it is the most likely of all outcomes for the short term. Truly, it could be argued that if the US government had rescheduled *Cannabis* many years ago, it theoretically could have kept it from becoming legalized at the State level, while also opening a path to DEA- and FDA- regulated prescriptions.

New scheduling category (Schedule VI/6)

Some reform advocates from the Schedule 6 Foundation are lobbying for Congress to create a *new* scheduling category [32]. The Schedule VI (6) Foundation praises their approach as "simple" and "elegant." Meanwhile it is criticized by some prominent administrative law scholars as "the worst *Cannabis* reform proposal yet" [33]. The basic idea? To amend the CSA to add a new scheduling category. Schedule VI – a category for substances that are "listed in Schedule I," while simultaneously "subject to some degree of abuse and has[ve] a currently accepted medical use in treatment" and/or are "restricted for sale to adults over the age of 21". Similar to the *de*scheduling outcomes, as projected "Schedule VI substances" would be subject exclusively to State regulation. However, marketing or product efficacy claims are already *known* to cross jurisdictional boundaries, especially via the internet, which can trigger the FDA's powers and jurisdiction on the matter. Problem not solved.

The three major flaws of the Schedule 6 approach, according to Pennington (2022), include but are not limited to: requiring *Cannabis* to remain in Schedule I to benefit from Schedule VI, no medical marijuana would qualify for Schedule VI placement (proposed legislation as written by the Schedule 6 Foundation), it fails to address international treaty issues, and "section 811(d)(1) of the CSA would render it useless." [33]

Any new scheduling category could also bring in a host of untoward and/or unforeseen issues, especially concerning existing legislative processes. Any legislative process would undoubtedly falter under the proposed language of the new amendment. As such, it is unlikely a new scheduling category could be created that does not follow the same logic as the existing schedules. The proposed amendment could be amended, arguably, to fit into the existing CSA, but this could bring about several other issues. For example, would all drugs and substances that are *not* scheduled then undergo an assessment, or be automatically included under the definition of a Schedule VI drug? This could open the door for scheduling unscheduled products related to alcohol, tobacco, OTC drugs like aspirin, and even prescription drugs like Epidiolex™. There are also substances considered less risky than certain *Cannabis* or hemp products that are currently scheduled. A chain reaction could be foreseen, one that results in multiple drugs being moved around amongst the various drug schedules. Indeed, this could likely create an absurd shake-up of the entire scheduling status of *every* substance in the CSA.

Otherwise, perhaps Schedule VI status could end up doing *nothing at all* – as it is a "self-defeating approach."

A new Federal agency focused on *Cannabis* and hemp

Conceptually, this seems a very likely scenario – certainly many years away from becoming a functioning reality if it were nestled within the FDA and, even more evidently, would perhaps take *decades* if it is a new agency *outside* the FDA. The FDA has hinted that some sort of dedicated group might be formed to address issues with CBD product safety, specifically, as well as D8-THC and/or other artificially derived hemp isomers. Questions abound though concerning the emergent structure, and authority, of any *de novo*-created agency – Who would work there? How it would be funded? Would there be a *Cannabis* regulatory body and a separate *hemp* regulatory body? Any dedicated agency with appropriate funding and a peaceful, scientific raison d'être might *sound* like a good thing, but these things seldom are, when focused on a singular product or industry. For instance, there is no single agency that focuses only on opioids . . . or insulin . . . or ibuprofen. Tobacco and alcohol do not have separate agencies – they are regulated across State and Federal organizations and somehow even got tangled up with firearms in the process. It is important perhaps that one agency or department is *not* the sole regulatory agency for *Cannabis* issues. For example, the FDA (as well as other agencies) has quietly stopped working on various programs, as Congress gradually decides to not appropriate funds, or when the Commissioner changes (usually every few years) and the focus of the agency changes accordingly.

As they have stated, "*The FDA looks forward to working with Congress to develop a cross-agency strategy for the regulation of these products to protect the public's health and safety.*" [23, 34] This strongly suggests that the existing regulatory framework used for foods and supplements is not appropriate for CBD. While any solution will surely involve a substantial amount of money (i.e., appropriations) to fund new regulatory efforts, the risk remains significant that a change in Commissioner or even members of the FDA's inner circle of *Cannabis* aficionados could *drastically* change the scope, focus, or direction of these efforts. Or, even more likely, Congress might not simply be *willing* to appropriate ample-enough funding for the FDA to undertake such a monumental effort. As mentioned earlier, the FDA cannot create new regulations without express permission or a request from Congress – and have you *seen* Congress lately? The FDA simply cannot just start regulating whatever product it wants, unless it is already covered under a law regulating FDA activity.[10]

Any new agency would likely work in tandem with FDA and other extant Federal agencies. This is similar to how the DEA, FDA, and HHS work together already – a new agency would just be another shrimp on the barbeque of Federal *Cannabis* regulation. Ultimately, such efforts will require massive amounts of time *and* funding from Congress. History shows that while this may be possible, the FDA has already abandoned several significantly less-aspirational programs because of a change in Commissioner or insufficient appropriations from Congress to fund the work that is needed.

Dietary supplements

Dietary supplements are considered a food, but are regulated separately from food, subject to specific and separate rules under the FDCA. *Cannabis* and hemp supplements navigate a very small footpath – they cannot legally contain THC or CBD, because both of

10 For example, "In the United States, federal laws are enacted by Congress. In order to make the laws work on a day-to-day level, Congress authorizes certain government agencies, such as FDA, to create regulations. A change in FDA's legal authority over cosmetics would require Congress to change the law." https://www.fda.gov/cosmetics/cosmetics-laws-regulations/fda-authority-over-cosmetics-how-cosmetics-are-not-fda-approved-are-fda-regulated#:~:text=In%20the%20United%20States%2C%20federal,Congress%20to%20change%20the%20law.

those are "approved drugs," Marinol^{TM} and Epidiolex^{TM}, respectively. "Approved drugs," as such, cannot be classified as a dietary supplement, which potentially leaves *some* room for other cannabinoidergic constituents. The CBD and hemp-extracted products we see marketed as dietary supplements, today, are done so *illegally,* and the FDA has explained as much in their recent regular unveilings of new warning letters [35]. To put it another way, enforcement action can be initiated based on *either* the disease claim *or* the presence of CBD. While a "Warning Letter" does not bind the FDA to take enforcement action, the recent endorsement of any new drug containing CBD might prompt the agency to act without said additional notification.

For this reason alone, it is near the top of the list of "unlikely possibilities" that either *Cannabis* or hemp will be authoritatively accepted as a "dietary supplement." Indeed, the FDA has been clear about regulating neither *Cannabis* nor cannabinoids as dietary supplements.[11] Alternatively, the FDA has suggested a separate agency be created to regulate only hemp products, and the FDA's longstanding position has been that CBD is excluded from the dietary supplement definition under section 201(ff) (3) (B) of the FDCA. There *does* remain a possibility that some form or specific product from hemp or *Cannabis* could be recognized as a dietary supplement, such as hemp protein powder or perhaps some form of animal feed, but this would assuredly be without THC, CBD, or other cannabinoid content as an active ingredient.[36–38] CBD is a molecule that has already been the subject of substantial clinical investigations, the existence of which have been made public (as well as being an article that is approved as a new drug), and CBD was not first marketed as a food or dietary supplement [39].

Finally, the FDA has very recently invited interested parties, including the public, to present any evidence that they think has a bearing on this issue [40]. The FDA's website dedicated to the regulation of *Cannabis* and *Cannabis*-derived products contains a statement *specifically* inviting interested parties to present the agency with such data. Warning letters have also stated that the recipient of the letter "may present FDA with any evidence that has bearing on this issue." The FDA states that as of 1/26/23, "they have not received or found any evidence that changes their position on CBD being excluded as a dietary supplement" [41].

11 "Congress defined the term "dietary supplement" in the Dietary Supplement Health and Education Act (DSHEA) of 1994. A dietary supplement is a product intended for ingestion that, among other requirements, contains a "dietary ingredient" intended to supplement the diet. The term "dietary ingredient" includes vitamins and minerals; herbs and other botanicals; amino acids; "dietary substances" that are part of the food supply, such as enzymes and live microbials (commonly referred to as "probiotics"); and concentrates, metabolites, constituents, extracts, or combinations of any dietary ingredient from the preceding categories." For more information see: https://www.fda.gov/food/information-consumers-using-dietary-supplements/questions-and-answers-dietary-supplements#:~:text=a%20mineral%2C,the%20total%20dietary%20intake%2C%20or

Conclusions

We should anticipate that some or all formal roles involving the creation of national policy and enforcement of *Cannabis* and the cannabinoids shall likely fall within the FDA's purview. Further, although purely intrastate activities *are* out of the FDA's jurisdiction, they are *not* out of the DEA's (see the Supreme Court case of Gonzales v. Raich) [42]. Also, the FDA's jurisdiction includes product components, not just finished products. Indeed, if FDA wanted to regulate a *Cannabis* brownie, it could do so if the flour and sugar and eggs and butter traveled through interstate commerce or internationally . . . even if the final product did not [43].

Alternative futures involving an entirely new federal agency may exist, but will almost certainly consist of *multiple* agencies and *multiple* departments working together, regardless. *Cannabis* and hemp are entirely unlikely to become less regulated than they already are. Indeed, existing GMP requirements or something emergently similar would certainly be required once new interstate commerce policies were solidified. Certain products may even require special risk management requirements, as seen with particular consumer products that already require refrigeration and/or additional warnings. **The most likely-to-succeed outcome would be a rescheduling, followed by cross-agency and industry collaboration.** Indeed, this seems entirely likely, because of both the historical and recent evolution of the FDA and its focus over the last decades, as well as the prevailing social climate of *Cannabis* acceptance. If *Cannabis* is moved to Schedule III, it will largely be a win for companies due to implementation of standard business deductions and emergent banking utilization – and who is it that greases the gears of Congress? Q.E.D.

> *Medicine is a social science, and politics is nothing more than medicine on a grand scale. – Rudolf Virchow, 1821–1902*

References

[1] Baum D Legalize It All: How to Win the War on Drugs. Harper's Magazine 2016.

[2] Downs D. The Science behind the DEA's Long War on Marijuana. Scientific American, 2016.

[3] Abuse, NC on M and D. Marihuana: A signal of misunderstanding. 1972.

[4] Weingarten G. Just What Was He Smoking?. Washington Post, 2002.

[5] 21 U.S. Code § 811 – Authority and criteria for classification of substances. n.d.

[6] Miller G After 4-year Delay, DEA will Review Dozens of Requests to Grow Marijuana for Research. Science 2020. https://doi.org/10.1126/science.abb9781.

[7] (DEA) DEA. 81 FR 53767 – Denial of Petition to Initiate Proceedings to Reschedule Marijuana 2016-17954 Docket No. DEA-427 2016.

[8] Marcu JP The DEA's Denial of Existing Medical Cannabis Research 2016. www.safeaccessnow.org/deareport_report (accessed 1AD).

[9] Jaeger K Biden Announces Mass Marijuana Pardons And Calls For Cannabis Scheduling Review. Marijuana Moment 2022. https://www.marijuanamoment.net/biden-announces-mass-marijuana-pardons-and-calls-for-cannabis-scheduling-review/.

[10] Jaeger K Top Federal Health Agency Says Marijuana Should Be Moved To Schedule III In Historic Recommendation To DEA 2023. https://www.marijuanamoment.net/top-federal-health-agency-says-marijuana-should-be-moved-to-schedule-iii-in-historic-recommendation-to-dea/ (accessed September 7, 2023).

[11] Mayl SL, Throckmorton D FDA Role in Regulation of Cannabis Products NIDA – Presentation 2019. https://www.fda.gov/media/128156/download (accessed July 4, 2023).

[12] Pruyn SA, Wang Q, Wu CG, Taylor CL. Quality Standards in State Programs Permitting Cannabis for Medical Uses. Cannabis Cannabinoid Res 2022. https://doi.org/10.1089/can.2021.0164.

[13] Gottlieb S Statement from FDA Commissioner Scott Gottlieb, M.D., on Signing of the Agriculture Improvement Act and the agency's regulation of products containing cannabis and cannabis-derived compounds (Dec. 20, 2018) [hereinafter Statement on Signing of the Agriculture Improvement Act] 2018. https://www.fda.gov/news-events/press-announcements/statement-fda-commissioner-scott-gottlieb-md-signing-agriculture-improvement-act-and-agencys (accessed July 4, 2023).

[14] Throckmorton DC FDA Role in Marijuana Regulation Presentation 2017. https://www.fda.gov/media/107242/download (accessed July 4, 2023).

[15] Simon TA, Simon JH, Heaning EG, Gomez-Caminero A, Marcu JP. Delta-8, a Cannabis-Derived Tetrahydrocannabinol Isomer: Evaluating Case Report Data in the Food and Drug Administration Adverse Event Reporting System (FAERS) Database. Drug Healthc Patient Saf 2023;15:25–38. https://doi.org/10.2147/dhps.s391857.

[16] Simon TA, Gomez A, Marcu J What Cannabis Related Data is Reported in the FDA FAERS Database. Cannabis Research Conference, 2021.

[17] Russo EB. Current Therapeutic Cannabis Controversies and Clinical Trial Design Issues. Front Pharmacol 2016;7:309. https://doi.org/10.3389/fphar.2016.00309.

[18] Russo E, Mathre ML, Byrne A, Velin R, Bach PJ, Sanchez-Ramos J, et al. Chronic Cannabis Use in the Compassionate Investigational New Drug Program. J Cannabis Ther 2002;2:3–57. https://doi.org/10.1300/j175v02n01_02.

[19] Mack J. Marijuana as Medicine? the Science beyond the Controversy. Washington (DC): National Academies Press (US), 2000. 11, LEGAL ISSUES. Available from: https://www.ncbi.nlm.nih.gov/books/NBK224398/. 2001.

[20] Kuromiya v. United States, 78 F. Supp. 2d 367 (E.D. Pa. 1999) 1999. https://law.justia.com/cases/federal/district-courts/FSupp2/78/367/2569063/.

[21] SB-1326 Cannabis: interstate agreements. https://leginfo.legislature.ca.gov/faces/billTextClient.xhtml?bill_id=202120220SB1326. n.d.

[22] Administration F& D. Food & Drug Administration Docket (FDA-2022-P-0600) Response Letter from FDA CFSAN to Natural Products Association 2023. https://www.regulations.gov/document/FDA-2022-P-0600-0005 (accessed July 4, 2023).

[23] Administration F& D. Food & Drug Administration Press Announcement: FDA Concludes that Existing Regulatory Frameworks for Foods and Supplements are Not Appropriate for Cannabidiol, Will Work with Congress on a New Way Forward 2023. https://www.fda.gov/news-events/press-announcements/fda-concludes-existing-regulatory-frameworks-foods-and-supplements-are-not-appropriate-cannabidiol (accessed July 4, 2023).

[24] Pennington S, Kline A, Mullin HB, Marcu J, Simon T Coalition for Cannabis Scheduling Reform Report 2023. https://schedulingreform.org.

[25] O'Connor SM, Lietzan E. The Surprising Reach of FDA Regulation of Cannabis, Even after Descheduling. Ssrn Electron J 2018;68:823–925. https://doi.org/10.2139/ssrn.3242870.

[26] Marcu J. Regulators Need to Rethink Restrictions on Cannabis Research. Nature 2019;572:S19–S19. https://doi.org/10.1038/d41586-019-02531-6.
[27] Jahan PM. Scheduling Cannabis: A Preparatory Document For FDA'S 8 Factor Analysis On Cannabis 2016. https://www.safeaccessnow.org/8_factor_analysis_on_cannabis.
[28] Marcu J. The Legalization of Cannabinoid Products and Standardizing Cannabis-Drug Development in the United States: A Brief Report. Dialogues Clin Neurosci 2020;22:289–293. https://doi.org/10.31887/dcns.2020.22.3/jmarcu.
[29] 21 U.S.C. 351 – Adulterated drugs and devices n.d. https://www.govinfo.gov/app/details/USCODE-2015-title21/USCODE-2015-title21-chap9-subchapV-partA-sec351 (accessed July 4, 2023).
[30] Knutson K Food Safety Lessons for Cannabis-Infused Edibles 2020;47–68. https://doi.org/10.1016/b978-0-12-819512-3.00004-3.
[31] Jameson LE, Conrow KD, Pinkhasova DV, Boulanger HL, Ha H, Jourabchian N, et al. Comparison of State-Level Regulations for Cannabis Contaminants and Implications for Public Health. Environ Health Persp 2022;130:097001. https://doi.org/10.1289/ehp11206.
[32] Schedule 6 Foundation n.d. https://schedule6.org (accessed July 4, 2023).
[33] Pennington S Schedule VI – The Worst Cannabis Reform Proposal Yet 2022. https://ondrugs.substack.com/p/schedule-vi?utm_source=profile&;utm_medium=reader2 (accessed July 4, 2023).
[34] Mead A. Legal and Regulatory Issues Governing Cannabis and Cannabis-Derived Products in the United States. Front Plant Sci 2019;10:697. https://doi.org/10.3389/fpls.2019.00697.
[35] Warning Letters and Test Results for Cannabidiol-Related Products 2023. https://www.fda.gov/news-events/public-health-focus/warning-letters-and-test-results-cannabidiol-related-products.
[36] Vispute MM, Sharma D, Mandal AB, Rokade JJ, Tyagi PK, Yadav AS. Effect of Dietary Supplementation of Hemp (Cannabis sativa) and Dill Seed (Anethum graveolens) on Performance, Serum Biochemicals and Gut Health of Broiler Chickens. J Anim Physiol An N 2019;17(59):533. https://doi.org/10.1111/jpn.13052.
[37] Mattila P, Mäkinen S, Eurola M, Jalava T, Pihlava J-M, Hellström J, et al. Nutritional Value of Commercial Protein-Rich Plant Products. Plant Food Hum Nutr 2018;73:108–115. https://doi.org/10.1007/s11130-018-0660-7.
[38] Zając M, Guzik P, Kulawik P, Tkaczewska J, Florkiewicz A, Migdał W The Quality of Pork Loaves with the Addition of Hemp Seeds, De-Hulled Hemp Seeds, Hemp Protein and Hemp Flour. Lwt 2019. https://doi.org/10.1016/j.lwt.2019.02.013.
[39] FDA Regulation of Cannabis and Cannabis-Derived Products, Including Cannabidiol (CBD) n.d. https://www.fda.gov/news-events/public-health-focus/fda-regulation-cannabis-and-cannabis-derived-products-including-cannabidiol-cbd (accessed July 4, 2023).
[40] Administration F& D. Food & Drug Administration Regulation of Cannabis and Cannabis-Derived Products, Including Cannabidiol (CBD) n.d. https://www.fda.gov/news-events/public-health-focus/fda-regulation-cannabis-and-cannabis-derived-products-including-cannabidiol-cbd (accessed July 4, 2023).
[41] FDA Issues Response to Three Citizen Petitions Related to CBD and Dietary Supplements 2023. https://www.fda.gov/food/cfsan-constituent-updates/fda-issues-response-three-citizen-petitions-related-cbd-and-dietary-supplements (accessed July 4, 2023).
[42] Gonzales v. Raich :: 545 U.S. 1 (2005) :: Justia US Supreme Court Center n.d. https://supreme.justia.com/cases/federal/us/545/1/ (accessed October 24, 2023).
[43] Marcu J Personal Communication with Howard Sklamberg 2023.

Sandra Carrillo

Cannabis innovations and regulation in Latin America

Cannabis legal status in Latin America

Cannabis was introduced in Latin America approximately in the sixteenth century, along with many other plants. As part of the increasing colonization process, the plant and its consumption were relatively common during the late sixteenth to the eighteenth centuries. In Brazil, it was mainly brought by slaves from today's Angola. For several centuries, it was grown and consumed primarily by the Afro-Brazilian and indigenous populations, having a solid presence in the rituals of some religions. In Mexico, hemp was cultivated since the early sixteenth century as an important agricultural commodity, while the medical uses of *Cannabis* have been known at least since the early eighteenth century. During the colonial period, hemp was cultivated to produce clothing and shipping accessories and used for ropes and other accessories in the mining industry, a crucial sector during the first centuries of colonialism. Hemp cultivation played an essential role in the industrial reconstruction of Chile after the War of Independence [1].

In the nineteenth century, *Cannabis* was introduced into modern European medicine, following research published by British doctors in the eastern colonies. *Cannabis* began to appear in pharmacopeias in several countries worldwide while its commercial market expanded, mainly as an anti-inflammatory, analgesic, antiemetic, and anticonvulsant [2].

Between the nineteenth and twentieth centuries, a trend toward prohibition and the persecution of those who used *Cannabis* and hemp gained force, mainly fostered by the new world-ruling classes and religious elite with their interests in the cotton industry and due to their conservative approach. Although relegated to the illegal market, *Cannabis* production flourished as Latin America became an international supplier of illicit drugs, suffering for years with a war on drugs and the consequences of unlawful trafficking, reinforcing the negative stigma attached to the plant [1].

The twenty-first century brought the beginning of the *Cannabis* comeback, with Uruguay legalizing its use in 2013 and others following the decriminalization and legalization trends. Since Uruguay's pioneering legalization of *Cannabis* in 2013, followed by Colombia in 2016, Latin America has been at the center of international interest due to its favorable climate, fertile soils, low labor costs, and sizeable domestic population [1].

Sandra Carrillo, Faculty of Medicine, University of Panama Scientific and Educational Medical *Cannabis* Program

https://doi.org/10.1515/9783110677508-003

The period between 2013 and 2018 could be considered the rebirth of the *Cannabis* and hemp industries in Latin America, with legalization in several countries. See Figure 1. An emerging sector started to gain shape and attracted the attention of international investors. Several companies established themselves in the region, mainly Canadian Licensed Producers (LPs), looking for emerging market opportunities [1].

Figure 1: Regulation of *Cannabis* for Medicinal Purposes in Latin America and the Caribbean.

This new context of regulations on *Cannabis* for medicinal purposes in Latin America could be a driving force for social justice. It could guarantee a therapeutic alternative for various medical conditions, reflecting scientific research and patients' positive experience using *Cannabis*-derived products for years. It could also be an opportunity to invest in repairing some of the social harms imposed by the long period of prohibition, which focused on the repression of vulnerable social groups involved in the production, commercialization, and consumption of *Cannabis*, leading to incarceration, exclusion, deepening inequalities, corruption, and the like [2].

Regulations for medicinal *Cannabis* should not reinforce the inequalities and violence historically associated with prohibitionist approaches but should enable the majority to access health care, especially the most vulnerable. Countries should make adequate provisions in their regulatory mechanisms that stress fair-trade principles that would protect such groups, especially traditional producers, and expand access to the range of *Cannabis* products for medicinal purposes [2].

This chapter will review some of the regulations affecting *Cannabis* in select countries throughout Latin America and the Caribbean.

The current legal situation in Argentina

In Argentina, the first law on medical *Cannabis*, approved in 2017 (#17,350), establishes a regulatory framework for the medical use and scientific research of this plant and its derivatives. In 2018, the first clinical research protocol in Latin America was approved at Hospital de Pediatria Garrahan (Buenos Aires, Argentina) to evaluate the efficacy and safety of cannabidiol (CBD) oil for treating pediatric patients with refractory epilepsy [3].

Advances in regulatory framework

The National Administration of Medicines, Food, and Medical Technology (ANMAT) permits the import of *Cannabis* oil and its derivatives when required by patients who present the pathologies contemplated in the program and have the appropriate medical indication. The provision is free for those who are incorporated into the program.

A voluntary national registry is created within the scope of the nation's Ministry of Health to authorize the registration of patients and the relatives of patients who have the pathologies included in the regulations and/or prescribed by doctors from public hospitals, and users of hemp oil and other derivatives of the *Cannabis* plant, with the protection of confidentiality of personal data [4].

Medical and scientific research (Regulation law 27350)

This law registers patients for treatment for case studies and patients in research protocols; the provision of *Cannabis* oil and its derivatives is free for those enrolled in the program and meet their requirements [5].

In July 2019, the promotion of the new regulation proposal of law 27,350 began. It aims to ensure access to the registry of medicinal specialties, public production, and controlled self-cultivation of *Cannabis*. Additionally, it seeks to expand medical specialties for *Cannabis* prescription, patient registration, and support for clinical research to generate evidence for its therapeutic indication in various conditions [4]

Reprocann program (Annex of law 27.350, 2020)

The Registry of the *Cannabis* Program (REPROCANN) is a database used for those who meet the conditions to access the controlled medical marijuana cultivation register for therapeutic and pain relief purposes. Its objective is to improve access to medical *Cannabis* for those with a medical indication, based on available scientific evidence – a product as a medicinal specialty, or compounded formulation [6].

The number of flowering plants allowed is a maximum of nine, the permitted surface of cultivation is up to six square meters, indoor cultivation is allowed, outdoor cultivation is not allowed, and transportation between one and six bottles of 30 mL and up to 40 g of dry flower is permitted [7].

Argentina, as of this writing, allows pharmacies to dispense *Cannabis* magistral preparations, allows access to patients under doctor's prescription for different medical conditions – patients will we able to register at the government program to grow at home or have an association grow for them. The government's objective is to facilitate and subsidize access to *Cannabis* derivatives to patients with a medical indication but this has not been implemented.

Current legal situation in Colombia

Colombia is a country that has suffered for many years from violence due to the war on drugs, social costs, health costs, and the environment. It is necessary to build a new drug policy, supported by evidence-based lessons learned from the previous experience of the last 50 years of the war on drugs. Colombia has advantages in its geographical position, climate, light conditions, sun conditions, and low labor and production costs [8].

In 1994, Colombian Constitutional Court determined that possession of *Cannabis* in amounts of personal doses was considered legal. In 2015, the Colombian Supreme Court allowed the cultivation for the personal use of up to 20 plants of *Cannabis*. In

2016, President Juan Manuel Santos sanctioned the law authorizing the medicinal use of *Cannabis* and its derivatives. For a timeline outlining the evolution of regulations, see Table 1.

Table 1: Timeline of Evolution and Advances in Regulatory Framework.

- **Law 1787-2016:** This law aims to create a regulatory framework that allows safe and informed access to the medical and scientific use of *Cannabis* and its derivatives in the Colombian national territory.
- **Decree 780-2016**: This decree is only about health and social protection sector regulation about the safe and informed access to *Cannabis* and the *Cannabis* plant.
- **Decree 613-2017:** By-Law 1787 is regulated and it is about safe and informed access to the medical and scientific use of *Cannabis*.

Psychoactive *Cannabis* plant cultivation license

Psychoactive *Cannabis* comes from the flowering or fruit-bearing tops of the *Cannabis* plant (except the seeds and leaves not attached to the tops), from which the resin has not been extracted, by whatever name they are designated, whose tetrahydrocannabinol (THC) content is equal to or greater than 1% by dry weight. Non-psychoactive *Cannabis*: non-psychoactive *Cannabis* is considered to be that which contains less than 1% THC [9].

In Colombia, there are different types of licenses for the manufacture of *Cannabis* derivatives. See Table 2 for a listing of these types of licenses.

Control of cannabinoids

In 2022, the Colombian Ministry of Health and Social Protection issued the Resolution 2808 of 2022. It ratified access to master formulas and medicines made with active ingredients of *Cannabis*, with coverage through the EPS (Entidades Promotoras de Salud; Health-promoting entities) and is financed with resources from the EPS Training Payment Unit [11].

For a person to have access to a preparation of master formulas, the preparation must comply with the standards defined by INVIMA, Good Practices in Elaboration, and a certification with extensions to *Cannabis*. The prescription must be made by a physician attached to the EPS network to which the patient is affiliated.

Manufacturing licensees must, at a minimum, determine through validated analytical methodologies, the THC, CBD, and CBN content of every *Cannabis* crop they receive and, in every batch produced [10]. Dispensing occurs through authorized pharmacies and the use of dry flowers for inhalation is not allowed for prescriptions. See Figure 2 for more information on *Cannabis* distribution throughout Colombia.

Table 2: Modalities of the License for the Manufacture of *Cannabis* Derivatives.

Type	Mode	Licensing body
Manufacturing license of *Cannabis* derivatives	– For national use – For scientific research – For export	National Institute for Drug and Food Surveillance INVIMA
Psychoactive *Cannabis* cultivation license	– For seed production and for sowing – For grain production – For the manufacture of derivatives – For scientific purposes – For storage – For final disposal	Ministry of Justice and Law
Nonpsychoactive *Cannabis* cultivation license	– For seed production and for sowing – For grain production – For the manufacture of derivatives – For scientific purposes – For storage – For final disposal	Ministry of Justice and Law
Seed license	– Scientific purposes – Marketing or delivery	Ministry of Justice and Law
Manufacturing license	– Derivatives for national use – From *Cannabis* derivatives for scientific research – *Cannabis* derivatives for import	Ministry of Health and Social Protection (Directorate of Medicines and Health Technologies)

Source: Article 2.8.11.2.1.2.

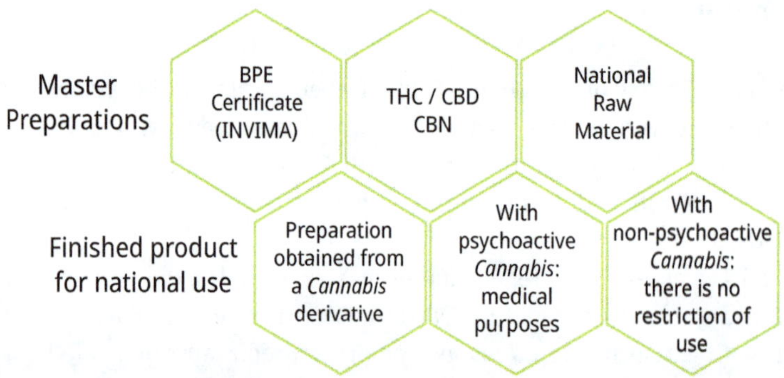

Figure 2: Products Containing *Cannabis* Derivatives for Medical Purposes and Their Distribution Nationwide.

The healthcare system will cover the *Cannabis*-based master preparations for patients with the following diagnosis: Fibromyalgia, refractory epilepsy, Dravet syndrome, Lennox Gastaut, cancer pain, chronic pain, noncancerous chronic pain, neuropathic pain, low back pain, insomnia, cancer cachexia, sleep, and appetite disturbances, among others [11].

Latin America, with the Colombian legislation, has one of the world's most advanced intellectual property regulations, when it comes to genetic intellectual property (IP), as the result of centuries of plant interbreeding, particularly in the second half of the twentieth century, as the country was a center for illegal cultivation [8].

In summary, Colombia has one of the most robust regulatory frameworks in Latin America, with a well-established system of licenses for cultivation, extraction, transformation, export, and research. Patients can access regulated preparations (Magistral Preparations) with doctors' prescriptions for more than a dozen indications. Dispensation of drugs is via pharmacies and is covered by the healthcare system.

Current legal situation in Peru

Timeline of evolution and advances in regulatory framework

On the twentieth of October 2017, members of the Peruvian Congress approved the medicinal use of *Cannabis* and finally, in November 2017, President Pedro Pablo Kuczynski officially published Law #30,681, which regulates the medicinal and therapeutic use of *Cannabis* and its derivatives.

This law aims to ensure the fundamental right to health, and enable exclusive access to *Cannabis* and its derivatives for medicinal and therapeutic purposes. It includes the registration of patient users, importers, and/or traders of *Cannabis* and its derivatives. Additionally, it establishes a registry for authorized research entities studying *Cannabis* and its derivatives for medicinal and therapeutic use. See Table 3 for the types of licenses in Peru [12].

Cannabis plant

Psychoactive *Cannabis* refers to products with a THC dry weight of more than 1%, excluding those that are combusted or smoked. This falls under the control process known as "*Cannabis* for medicinal use." On the other hand, nonpsychoactive *Cannabis*, including its parts and derivatives with less than 1% THC dry weight, is categorized as noncontrolled substances under the regulation known as "hemp."

Table 3: Types of *Cannabis* Licenses for Various Purposes.

Type	Applicant	Licensing body
Import license	– Public entities – Drugstores – Pharmacy	MINSA(Ministry of Health)/DIGEMID(Direction of Medications and Drugs) MINSA/DIGEMID
Production license	– Universities and research entities – Public institutions – Laboratories	
Research license	– Universities and health research entities – Universities and agricultural research institutions	MINSA/INS(National Institute of Health) MINAGRI/INIA

Only medical surgeons can prescribe *Cannabis* derivatives for medicinal use and therapeutic purposes. Medical doctors can prescribe the necessary conditions under their medical criteria and responsibility [12, 13].

Cannabis derivatives for medicinal use

For the purpose of this regulation, any compound, mixture, preparation, or product derived from *Cannabis* intended for therapeutic use includes: herbal medicine (Sativex), pharmaceutical preparation (Master Formulas), pharmaceutical product, and natural product for health use (Epifractan). Medical prescriptions for *Cannabis* are categorized as special prescriptions for *Cannabis* derivatives used for medicinal purposes, or simple prescriptions, limited to CBD in its formulation. The dispensing of these prescriptions is carried out by pharmacies in public or private sector establishments holding the appropriate license [13]. Possession of *Cannabis* for immediate consumption is not punishable in an amount that does not exceed eight grams of marijuana or two grams of its derivatives.

Possession of *Cannabis* and its derivatives for medicinal and therapeutic purposes will not be punishable either, provided that the amount is necessary for the treatment of the patient registered with the Ministry of Health and supervised by the National Institute of Health and the General Directorate of Medicines, Supplies, and Drugs DIGEMID, or of a third party that is under its care or guardianship, or for investigation according to the laws on the matter and the provisions established by the governing body [12].

In summary, Peru has a progressive regulatory framework for medical *Cannabis*, allowing the cultivation, production, import, research and sale of *Cannabis* for medical purposes; patients with a qualifying medical condition can access *Cannabis*-based products under a doctor's prescription, and products are sold by pharmacies.

Current legal situation in Chile

Timeline of evolution and advances in regulatory framework

Decree 404 and Decree 405 grant regulatory power to the Institute of Public Health over crop authorizations for medical or scientific research purposes [15, 16]. Medical *Cannabis* has been legal in Chile since 2017 [14].

Law 20,000, which addresses the illicit trafficking of narcotic drugs and psychotropic substances, undergoes modifications alongside the Sanitary Code. These modifications aim to legalize the private self-cultivation of *Cannabis* and decriminalize its sale and cultivation for therapeutic purposes [14]. Furthermore, Decree 867 provides the specific requirements for individuals interested in establishing *Cannabis* cultivation in Chile [14]. Additionally, through Decree 84, the Chilean government acknowledges and legalizes the medicinal use of *Cannabis*. Pharmaceutical products containing *Cannabis*, *Cannabis* resins, extracts, and tinctures can be dispensed through pharmacies or laboratories, with medical prescriptions, subject to controls as stipulated by the Law. The import, export, transit, extraction, production, manufacture, division, preparation, distribution, transport, transfer in any capacity, sale, possession, and possession of *Cannabis*, *Cannabis* resin, extracts, and tinctures of *Cannabis*, will be prohibited in the national territory. Notwithstanding the foregoing, in qualified cases and for scientific research purposes, the Chilean Institute of Public Health may authorize using these substances under the conditions determined by the corresponding resolution.

Modification to the approved law (March 14/23): "The cultivation of plant species of the *Cannabis* genus for the attention of a medical treatment will be understood to be justified, with the presentation of the prescription issued for that purpose by a treating surgeon, who must indicate the diagnosis of the disease, its treatment, and duration, in addition to the form of administration of the *Cannabis*, which may not be by combustion." [15, 16].

In summary, the Chilean regulatory framework allows the cultivation, production, and distribution of *Cannabis*-based medicines by authorized companies; patients have access to *Cannabis*-based medication with a doctor's prescription under certain qualifying conditions and can purchase *Cannabis*-based drugs from authorized pharmacies.

Current legal situation in Uruguay

Uruguay is a pioneer in legalization and regulation. It became the first country in the world to legalize *Cannabis* when President Jose Mujica signed the bill into law in December 2013. Currently, in the country, the cultivation, commercialization, and consumption of *Cannabis* are allowed, with restrictions regarding the amount that can be acquired and cultivated, but not in terms of its medicinal or recreational purposes [17]. See Figure 3 for the timeline of regulatory evolution in Uruguay.

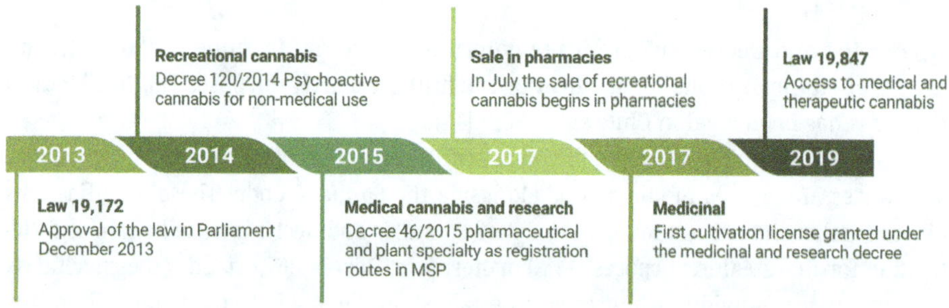

Figure 3: Evolution of *Cannabis* Regulation in Uruguay. Comparative Milestones of Recreational and Medicinal Markets.

Timeline of evolution and advances in regulatory framework

Law 19,172 refers to regulating the production, distribution, and sale of *Cannabis*. The standard is part of a policy aimed at minimizing risks and reducing damage, promoting information, education, and prevention of the problematic use of this product [18].

The Uruguayan government is even taking the final steps to regulate medicinal formulas based on *Cannabis* extracts almost three years after the enactment of the law implementing the National Medical *Cannabis* Program in Uruguay and nearly ten years after its legalization [17].

The master formulas prescribed by doctors and elaborated by pharmaceutical chemists are one of the ways for Uruguayans to have safe access to medical *Cannabis* and a tool for the doctor to formulate precisely what the patient needs and incorporate *Cannabis* into the medical practice. Healthcare professionals can prescribe *Cannabis* based on their clinical judgment [17].

In summary, Uruguay has a legal frame work that allows for recreational and medical use, *Cannabis* cultivation, and industrial hemp production. Regulations aim to find a balance between individual freedom, public health, and responsible use.

Current legal situation in Mexico

In 2017, the Mexican President Enrique Peña Nieto issued a mandate to legalize medical *Cannabis* in the country, following the senate vote in December, 2016. The 2017 decree amended the country's General Health Law and Federal Penal Code to regulate the medicinal use of pharmacological derivatives of *Cannabis* sativa, which also stated that *Cannabis* derivatives with less than 1% THC that have industrial uses may be marketed, exported, and imported in compliance with the requirements established in the health regulations [28]. Planting permits for authorized *Cannabis* species or varieties will be granted for planting, cultivation and harvesting activities, as well as for health research and manufacture of pharmacological derivatives and medicines [19].

To import or export raw materials, pharmacological derivatives, and *Cannabis* medicines, it is necessary to obtain a prior sanitary permit for import or export. This permit is granted by SADER or COFEPRIS (Federal Commission for the Protection against Sanitary Risks) within the scope of their respective authorities [19]. Professionals seeking to obtain a barcode for a specific prescription of *Cannabis* medicines must submit a request using the prescribed format issued by COFEPRIS [19]. Similarly, individuals interested in conducting research for pharmaceutical development and production, including clinical trials, must obtain authorization for the investigation protocol from COFEPRIS [19]. Under the 2017 General Health Law, patients with certain conditions can access low-THC *Cannabis* with a prescription from a doctor, however no legal structure has been put in place allowing for legal sale, nor are there provisions for commercial cultivation and production.

In summary, in Mexico, the regulatory framework is in a state of transition, while there has been progress for both medical and recreational *Cannabis*. State and federal penal laws remain in place, pending action by Mexican Congress. The medical *Cannabis* rules were published in January of 2021, allowing doctors to prescribe *Cannabis*-based medicinal products, but as of mid-2023, there is no wide access to products for patients, and importation is the only way to have access.

Current legal situation in Panamá

Law No. 242, approved on October 13, 2021, regulates the medical and therapeutic use of *Cannabis* in Panama [20]. The permitted purposes of *Cannabis* within the national territory are limited to medical, therapeutic, veterinary, scientific, and research applications [13]. Importing, exporting, re-exporting, and manufacturing *Cannabis* will require an operational license [20].

In order to maintain control and ensure patient safety, a record of *Cannabis* users with a medical certificate will be maintained. Prohibitions include home or internet marketing, selling in health food stores and doctor's offices, as well as promotion and advertising. The sale of *Cannabis* will only be permitted in authorized and certified pharmaceutical establishments [20]. The local production of *Cannabis* will be subject to strict monitoring and surveillance by the Ministry of Health and Social Protection (MINSA). Authorized laboratories must meet ISO standards [20].

Two types of local manufacturing licenses are available: the manufacture of *Cannabis* derivatives for national use or export, and the manufacture of *Cannabis* derivatives for scientific research. These licenses, granted for a period of ten years, can only be cancelled or suspended in accordance with the causes established by law. Moreover, these licenses are exclusively awarded to legal entities [20].

More than 23 medical conditions are included in the protocols for the therapeutic use of medical *Cannabis*, and a registry for patients using medical *Cannabis* will be created. This will have the doctor's indications for the patient. Every doctor with an identity in Panama can prescribe, after certification in *Cannabis* prescription, with the approval of the Ministry of Health. The medical prescription must comply with the requirements established by the Ministry of Health. The administration routes include oral presentation, suppositories, and flowers [21].

In summary, Panama has a very complete regulatory framework for medical *Cannabis* that is waiting to be implemented.

Current legal situation in Brazil

Timeline of evolution and advances in regulatory framework

In Brazil, the species *C. sativa* is included in the list of banned plants that can originate narcotic and/or psychotropic substances. In 2006, Law #11.343/2006 provided for the possibility of government authorization of the species cultivation, exclusively for medicinal or scientific purposes, under the state's surveillance, which is coherent with what is already foreseen in the international treaties in force [22].

Since 2015, there have been some notable advances toward allowing the medicinal use of the species in the country. With the publication of Brazilian National Health Surveillance Agency (ANVISA) RDC N° 03/2015, cannabidiol was moved from List "F" (Banned substances in Brazil) to list "C1" (list of other substances subject to exceptional control) of the *Portaria* N° 344/1998. In the same year, importing CBD-containing products for medical use was simplified. This was motivated mainly by reports of CBD's effectiveness in treating refractory epilepsy in children [22] In 2017, Mevatyl® (Nabiximols) became the first *Cannabis*-based medicine to receive marketing authorization in Brazil, with the therapeutic indication for MS-related spasticity [22].

The ANVISA allows the sale of *Cannabis* products, through a regulation, with two types of prescriptions. The first refers to compounds with a percentage greater than 0.2% THC for terminal patients, while the second refers to products with *Cannabis* less than 0.2% THC. *Cannabis* products will be authorized for oral or nasal use only and must be of pharmaceutical quality for human use [24]. Commercialization in the country will be carried out exclusively in pharmacies of manipulation (master) or drugstores, and with medical prescriptions.

Export opportunities for prioritized products

Brazil has more than 200 million people with great internal potential for users of this type of product. The country is estimated to have the highest concentration of potential consumers in the region, and a domestic market, which may be contemplated to be 3.4 million patients [25]. The "in-nature" plant continues to be proscribed; that is, its cultivation in Brazil is still a crime, and Phytocannabinoids (CBD and THC) are on the list of substances subject to special control, but with the possibility of medicinal use (Ordinance 344 of the Ministry of Health) [26, 27].

Medical doctors will be responsible for the prescription of *Cannabis* products for clinical conditions, in the absence of therapeutic alternatives [26]. Judicial decisions ensure that patients and patient associations have access to the plant to produce the medicine [26, 27].

Current rules from Brazil's ANVISA already allow for the importation and prescription sale of *Cannabis*-derived medicines, registered as pharmaceuticals and those with health authorization that do not require clinical trials under a regulation that went into effect in March 2020. That rule also provides a "compassionate use" license, allowing patients to import *Cannabis* products individually [23].

In summary, in Brazil, medical *Cannabis* is permitted only for medical use, with authorization from the ANVISA. Medical doctors must prescribe *Cannabis*-based medications for clinical conditions without other therapeutic alternatives; the products may be imported; local production is not allowed.

Conclusions

The evolution of medical *Cannabis* regulations in Latin America and the Caribbean has attracted enormous interest from large *Cannabis* corporations, especially those based in the US and Canada. These corporations have identified countries in the region as potential consumer markets for their products.

Some key countries, such as Colombia, have also been seen as potential centers of *Cannabis* production for the global market because of their favorable economic and environmental factors. The way these large *Cannabis* corporations access the Latin America Caribbean market is through partnerships of different types with local companies. These countries' companies and governments consider the large corporations' direct investments, knowledge, and technology transfer, indispensable.

Dependence on investment by these corporations creates massive power asymmetries in defining the conditions for their insertion in emerging markets. It makes it possible for these transnational corporate partnerships to extract wealth at the expense of local development and the livelihoods of small and medium farmers and other vulnerable groups historically engaged with the illicit *Cannabis* market.

In addition, the environmental threat arising from large-scale *Cannabis* production, the multiplication of patents that seek to monopolize niches in the new markets, and the high requirements for certification of good practices pose challenges for those interested in more democratic and fair *Cannabis* regulations.

Overall, the corporate capture of the new *Cannabis* regulations in Latin America has led to valorizing a pro-profit approach that reinforces inequalities. Medicinal *Cannabis* regulations should not strengthen the inequalities and violence created by various forms of prohibition throughout history. Indeed, they should help the majority of the population to access health care, especially the most vulnerable.

Bibliography

[1] Partners P. The Latin America and Caribbean Cannabis Report. Second Edition. 2020.

[2] https://www.tni.org/files/2023-01/policybrief_53_web.pdf (Accessed March 22, 2023)

[3] Cáceres Guido P, et al. Medicinal Cannabis in Latin America: History, the Current State of Regulation, and the Pharmacist's Role in a New Clinical Experience with Cannabidiol Oil. J Am Pharm Assoc: JAPhA 2020;60(1):212–215. doi: 10.1016/j.japh.2019.09.012.

[4] https://www.argentina.gob.ar/noticias/trabajo-comun-para-el-control-y-desarrollo-de-canamo-industrial-y-el-cannabis-medicinal. (Accessed March 22, 2023)

[5] https://www.argentina.gob.ar/normativa/nacional/decreto-738-2017-279831/texto. (Accessed March 22, 2023)

[6] https://www.argentina.gob.ar/salud/cannabis-medicinal/reprocann. (Accessed March 22, 2023)

[7] https://www.boletinoficial.gob.ar/detalleAviso/primera/241814/20210312. (Accessed March 22, 2023)

[8] Partners P. The Latin America and Caribbean Cannabis Report. Second Edition. 2020.

[9] https://www.minsalud.gov.co/sites/rid/Lists/BibliotecaDigital/RIDE/DE/DIJ/decreto-613-de-2017.pdf. (Accessed March 22, 2023)

[10] https://www.minsalud.gov.co/Normatividad_Nuevo/Decreto%20613%20de%202017.pdf. (Accessed March 22, 2023)

[11] https://www.invima.gov.co/cannabis. (Accessed March 22, 2023)

[12] https://cdn.www.gob.pe/uploads/document/file/297302/Decreto_Supremo_N%C2%BA_005-2019-SA.PDF. (Accessed March 22, 2023)

[13] Decreto Supremo No 005-2019-SA que aprueba el Reglamento de la Ley No 30681, Ley que regula el uso medicinal y terapéutico del Cannabis y sus derivados. [en línea]. Lima: Congreso de la República; 2019. Disponible en: https://busquedas.elperuano.pe/normaslegales/decreto-supremo-que-aprueba-el-reglamento-de-la-ley-n-30681-decreto-supremo-n-005-2019-sa-1744045-2/. Consultado 10 Nov 2019. (Accessed March 24, 2023)

[14] https://www.sag.gob.cl/ambitos-de-accion/solicitud-siembra-cannabis. (Accessed March 24, 2023)

[15] https://www.bcn.cl/leychile/navegar?idNorma=235507. (Accessed March 24, 2023)

[16] Sánchez S Dr, Bustos SS. A surgeon from the University of Chile. MPH, MBA, PhD in socio-sanitary sciences and medical humanities from the Complutense University of Madrid. https://www.bcn.cl/leychile/navegar?idNorma=235507&idParte=8652270&idVersion=2005-02-16). (Accessed March 25, 2023)

[17] https://www.meritas.org/insight/article/Historical-Overview-on-the-evolution-and-legislation-of-cannabis-in-latin-america. (Accessed March 24, 2023)

[18] https://medios.presidencia.gub.uy/jm_portal/2014/noticias/NO_M871/reglamentacion-ley19172.pdf. (Accessed March 24, 2023)
[19] http://www.dof.gob.mx/nota_detalle.php?codigo=5609709&fecha=12/01/2021 (Accessed March 25, 2023)
[20] https://www.gacetaoficial.gob.pa/pdfTemp/29398_B/87966.pdf (Accessed March 25, 2023)
[21] https://www.gacetaoficial.gob.pa/pdfTemp/29612_A/GacetaNo_29612a_20220901.pdf. (Accessed March 24, 2023)
[22] de Souza MR, Henriques AT, Limberger RP. Medical *Cannabis* Regulation: An Overview of Models Worldwide with Emphasis on the Brazilian Scenario. J *Cannabis* Res 2022 Jun 16;4(1):33. doi: 10.1186/s42238-022-00142-z. PMID: 35710586; PMCID: PMC9202119.
[23] https://www.meritas.org/insight/article/Historical-Overview-on-the-evolution-and-legislation-of-cannabis-in-latin-america. (Accessed March 25, 2023)
[24] https://www-in-gov-br.translate.goog/en/web/dou/-/resolucao-rdc-n-660-de-30-de-marco-de-2022-389908959?_x_tr_sl=pt&_x_tr_tl=es&_x_tr_hl=es-419&_x_tr_pto=sc. (Accessed March 25, 2023)
[25] https://www.pgrweb.go.cr/scij/Busqueda/Normativa/Normas/nrm_texto_completo.aspx?param1=NRTC&nValor1=1&nValor2=96518&nValor3=130412&strTipM=TC. (Accessed March 25, 2023)
[26] https://abracro.org.br/pdfs/04_RA_RDC38-2013_MANUAL_acesso%20exp_uso%20compas_JUL.pdf.
[27] Navarro LS Dr. Advogado – OAB/SP 163621 Especialista em Direito Médico Dr. Paula Dall´Stella, Médica – CRM/SP 128122 MD, Integrative Medicine, Hospital Sirio Libanes.
[28] https://www.dof.gob.mx/nota_detalle.php?codigo=5487335&fecha=19/06/2017#gsc.tab=0

Experts in the Field Interview: Gillian Schauer, PhD

For video of this interview scan here

Biography

Gillian Schauer, PhD, MPH was elected CANNRA's first Executive Director in 2021, and served as a Senior Advisor to CANNRA before the association's founding. Dr. Schauer has worked in public health and policy for nearly two decades, and has a decade of experience working with federal and state agencies on *Cannabis* policy, data monitoring, and research translation. Through her federal work, she founded and led the Multi-State Collaborative for *Cannabis* and Public Health from 2014–2021 to foster learnings and cultivate best practices for public health across states with legal *Cannabis*. Prior to working on *Cannabis* policy, Dr. Schauer worked on tobacco control policy. She was the Senior Editor on the 2020 Surgeon General's Report on Smoking Cessation. Dr. Schauer is an affiliate researcher at the University of Washington and has more than 70 peer-reviewed research publications on *Cannabis* and other substances. She has a PhD from Emory University, a Master of Public Health from University of Washington, and a Bachelors of Science from Northwestern University.

Jahan Marcu:

Hello, I am Jahan Marcu.

Andrew Peterson:

And I am Dr. Andrew Peterson, and we are the editors of the *Cannabis* Innovation Series. We would like to welcome you to another Interview with the Experts in the Field interview. This volume of the series is focused on regulated *Cannabis* and hemp market navigation, examining business, public health, and research challenges. We are here today to talk about regulatory affairs, public health, and research challenges. With us today is Dr. Gillian Schauer, the executive director of the *Cannabis* Regulators Association or CANNRA. CANNRA is a nonpartisan association of government officials involved in *Cannabis* regulation.

Jahan Marcu:

We are here today to talk about regulatory affairs, public health, and research challenges. With us today is Dr. Gillian Schauer, the executive director of the *Cannabis* Regulators Association or CANNRA. CANNRA is a nonpartisan association of government officials involved in *Cannabis* regulation.

Andrew Peterson:
So, Dr. Schauer, if you don't mind, just tell us a little bit about what is CANNRA and why is it important to have an association like this?

Gillian Schauer (CANNRA):
Yeah, so CANNRA is an association of government officials. We bring together government agencies that are regulating *Cannabis*, implementing *Cannabis* policy across 45 US states and territories and internationally as well. So, we have Canada as a member and have regular discussions with other countries. I think it's been really important. We were founded three years ago in 2020, but we were the offshoot of really . . . informal gatherings that regulators were having since the beginning of legalization, really, in Colorado and Washington. I think in the absence of the federal government being engaged in the U.S., states really started to realize, we are the policy lab, and we need to learn from each other. We need to be sharing what's working and what's not working. We need to have a peer-to-peer community where we can talk about challenges together. And so that started informally, and that's what continued to be our focus. How can we help convene government agencies to talk about the challenges? How can we work to identify best practices? How can we share resources when that's needed? States are not generally as well-resourced as the federal government. So, being able to just share lessons learned and grow in policy together, I think has been the most valuable part of coming together. And now internationally, as we've seen policy change, there's an interest, both in international governments to learn from US states and vice versa. Some international governments, as I understand you're covering in the book, have done things very differently than we have in the United States. And so, US states are equally interested in learning. So, it's been exciting to broaden that exchange outside of the US and also include international governments in our community as well.

Andrew Peterson:
I want to just follow up. Just three years ago, you started and you've grown to 45 states plus international.

Gillian Schauer (CANNRA):
Yes.

Andrew Peterson:

Wow, that's an amazing growth.

Gillian Schauer (CANNRA):
There was a demand and we are still very small staff. We are . . . it's really just me on the staff side. We've got lots of volunteer engagement from our members who are in the field doing the work. But I think there's really been a demand for it. And I have seen people who are already overworked in their day jobs give so much to CANNRA because they're excited about the opportunity to learn and to influence policy on the

ground from what they're seeing, in meaningful ways. This is such a critical time for *Cannabis* regulatory policy. We're really at the beginning of something and I think state regulators have a front row seat to see how this works and to make tweaks and changes and learn from those.

Jahan Marcu:
Well, speaking of tweaks and changes and growth, you know, it seems like every month, week or day-and-a-half, something new is happening in the *Cannabis* space. So how is CANNRA preparing for growth in the *Cannabis* industry? Things like new products, the rollout of adult use legislation, and even, potentially, federal guidelines.

Gillian Schauer (CANNRA):
Yeah, so we have seen this be a very fast paced arena, which I think, you know, some of our regulators have come from other regulatory domains and one of their first comments is always wow, this is moving so much faster than whatever I was regulating before. So again, the importance of having that exchange. We work through committees so we have a dozen different special committees that are co-chaired by government staff who are doing that work in their state, territory or international government agency, and those committees are really focused on tracking these fast-paced changes and talking about what they mean. So, we've got a lab testing and product safety committee, and they're focused on what new products we are seeing. Do we have the right methods to test for what we need to in those products? And do we have an ability to test for this new cannabinoid that we're seeing? We have a committee focused on interstate commerce and what will happen to state programs in the eventual – a federal legalization scenario. We have groups thinking about the transition from medical to adult use, which we've now seen many states go through. So, we have a lot of good state case examples of, you know; this helped my medical program when we were transitioning, or I wouldn't have done this the same because of these reasons. So, we really bring that to our members to talk about with each other and track. And then there're some issues that are salient enough that we take them on at a . . . broader organizational association level. So interstate commerce is one of those where, the work we need to do from a state perspective goes above and beyond what any one state could do. We really need to pull states together to think about how do you harmonize? What federal minimum standards would we want to see? How can regulators be engaged in setting those? We've also had a much more concerted effort this last year to engage with organizations that are working to make standards for *Cannabis* – standard development organizations. And traditionally, regulators who are state government officials have had some challenges getting engaged with those organizations, in part because the nature of the discussion about how do you set a standard and what's the best practice that gets into territory where many state government officials might feel, gosh, I need to clear this through the governor's office to have permission to share this or say this because it is effectively making new policy, and states have been very careful not to want to engage in discussions where it might

telegraph some big shift in state policy when really it's just an expert saying, well, if we were doing it again, we'd probably want to do it this way. So I think we've found some ways to have communication channels that are open with the groups that are making standards so that regulators can be more engaged without the pressure of them feeling like their telegraphic new policy or need to have clearance for something that they're going to say that's really their opinion as a subject matter expert. So, I think it's really important to have that standard development work. Regulators have to be at the table for that. And we have to be thinking not just one year ahead, but five or 10 years ahead to try to figure out how we're all going to work together. And these different state systems are eventually going to work in a framework where we have some federal overlay for federal legalization.

Andrew Peterson:
Speaking of federal legalization, Dr. Schauer, just last week, the HSS had recommended that *Cannabis* be placed in the Schedule III category. How might this affect your work? I know you folks are doing a lot as it is, but that's probably going to boggle your mind to begin with. And then even if you could think about how does that affect intellectual property and product development too.

Gillian Schauer (CANNRA):
I think we are paying very close attention to the discussions about rescheduling. We knew it was happening. In fact, the *Cannabis* Regulators Association, we put a call out in early January to ask our state members to send any data in that they thought might be useful to FDA in their rescheduling review. As you may have covered on prior videos, the rescheduling process that was called for is really a scientific review. And we knew right away that rescheduling process, if it was just based on peer-reviewed literature, would be missing a lot that's out there because we've had such a challenge in conducting *Cannabis* research here in the US. And state programs have a lot of data and research that they aren't able, for a variety of reasons, to publish through peer review channels. They don't have capacity or time or staff, or in some cases, it's brand new data. And so we were able to have a number of states share data directly with FDA to weigh in on their scientific review for that process. So, we knew rescheduling was being discussed. I think, you know, whether it's schedule II or III or IV or V, some things won't stay change in state programs. State programs would still be under any scheduling of marijuana federally illegal because they are operating outside the bounds of what the Controlled Substances Act calls for. But there are some things that could change, notably tax liability and the ability to deduct things for the industry, which I think states are watching very closely. There are also some important things that wouldn't change, like the criminality of marijuana that states are also watching closely and concerned with. So, we've been in a real listen mode, working with stakeholders at the national level to understand. What are their perspectives on different scheduling of *Cannabis* and scheduling versus descheduling and what would they like to see from their regulator? And I think we're starting to get some real tangible sug-

gestions of areas where regulators can provide increased TA for certain things, can continue to push state programs forward in spite of what might be uncertainty at the federal level. That's a big thing that we hear is for businesses, they depend on having some certainty in regulation and a lot of what's happened at the federal level could create uncertainty for a number of months or potentially years, and that can be hard for a business. So, I think we're encouraging members to stay focused on the mission that they've set forward in their state or jurisdiction. And we will work as a community to sort of figure out what's changing and evolving at the federal level and what the implications might be, but having that consistency for their state program, I think will be very important in the short term.

Andrew Peterson:
And I bet that the states appreciate having that kind of focus on and helping them focus on what their mission is within the state and recognizing that CANNRA is the one that's out there, kind of advocating for and pulling all of this stuff together. That might give them a little bit of that sense of stability that they need in these uncertain times. So kudos to you and your organization for being able to pull that through in this crazy time because as you said, even if they do reschedule it to a schedule three, that creates a whole bunch of other uncertainties and the illegality of current systems. For you to be able to do that in your organization, be able to do that, it's really a bonus, I think, for the . . . And that's much needed, so much appreciated.

Gillian Schauer (CANNRA):
We hope so.

Jahan Marcu:
It was really great to hear that states have the data. Maybe the data is all around us, but it's not evenly distributed in terms of access and publishing. I've got to say, if there was a job posting to be a science writer to write up the state data, I would jump on that. You mentioned that it's difficult for regulators to find time to crunch that data, write up that data, then go through the peer review process because regulators such as yourself are also providing written testimony and testifying before members of Congress and elected officials. You recently responded to the RFI about hemp. I definitely would like to get a sense of any insights you can provide us about how was that received by elected officials and maybe even the general public. Was it better than you thought? Was it . . . as well as you thought it would be. Did you feel like you really got their ear or do we need to clean out their ears and shout the message even louder?

Gillian Schauer (CANNRA):
Well, I do think this has been the summer of hemp. We've had a lot of opportunity to give input to Congress and it's a very important issue. I don't know how in depth you've covered it already, but it is one of the most pressing regulatory issues for states right now. We effectively have federal legalization that is so broad for hemp that

we're seeing products that . . . not only are the same as what we see in adult use and medical use *Cannabis* markets, which are, again, federally illegal, we are also seeing products that extend far beyond what any state marijuana program would legalize. And so there's a very acute threat to consumer safety and public health in addition to just the need to provide regulation. I think most consumers, when they purchase something at a store, they assume that product is well-regulated. And unfortunately, that is just not the case with cannabinoid hemp products today. So, it's been a big focus. Every single regulatory agency in our membership has consumer safety at the forefront of their regulatory work. That's why they regulate *Cannabis*. So, we'd like to see that extend to hemp as well. And that's why we've had such a focus on it. So, there've been a number of things we've gotten to be engaged in over the summer. The House Oversight Committee in Congress had a hearing about it, and we were . . . very happy to see that they did include CANNRA. They included a regulator as part of that hearing. And then shortly thereafter, the Energy and Commerce Committee and the Senate Help Committee put out this request for information about cannabinoid hemp and regulation. And if anybody understands regulation of cannabinoids, it's our members. They are the ones who are doing the work on the ground right now. And I think we have a lot to share with Congress because of what they've learned. So it was a welcome opportunity. I think we have a couple things on the horizon that we hope will yield change in how hemp is currently being regulated. We really hope the farm bill will have some modifications to address some of what we're seeing in terms of the lack of regulation and the breadth of products, but it remains to be seen. So, I think our goal whenever we have the opportunity to engage with Congress is really helping elected officials to meet their regulator. It's surprising how few elected officials know that they, in some cases, even have the *Cannabis* regulator. So, I think that's been one of the biggest pieces of our work. We're not a lobbying association. We're an association of government officials. We try to speak with one voice for the regulators, but I think maybe the most impactful thing we can do is just any outreach we get from Congress, immediately connecting them in with their regulators so they can hear firsthand, this is what this issue looks like on the ground in my state, in my jurisdiction. These are pictures of the products we're seeing, these are health events that have happened in my jurisdiction, whatever it may be. So, we hope to continue to do more of that into the fall and be a resource in whatever way we can for elected officials. I think we have a unique space that we occupy. We don't have a horse in the race, so to speak. We are the government officials who are regulating the industry. So, we don't have a vested financial interest on one side of an issue or another. We hope that makes us a group that's approachable for education to Congress. When they know that we are a nonpartisan, neutral party with direct on-the-ground experience regulating, we hope that that's a resource that they'll take advantage of.

Jahan Marcu:
Absolutely. I thought that was a fascinating statement. The summer of hemp.

Andrew Peterson:
[Laughs]

Jahan Marcu:
Do you think we'll have a few more summers where hemp is the focus of these discussions or . . . ?

Gillian Schauer (CANNRA):
Oh, I hope not, but I think it is going to be very difficult federally, because what effectively has happened through the really broad federal legalization of hemp in 2018 is we have basically federally legalized marijuana. We've just called it hemp and it has no regulation. And so I think that's challenging. Congress has not agreed on what to do with marijuana. And so the idea that they may have already inadvertently legalized marijuana under the name of hemp is it doesn't sit well with everybody. So, I think we will continue to see state after state try to protect consumer safety within the bounds of their state in any way that they can. But without federal action, there will still be some big gaps. So, state law, you know, doesn't easily extend to say online marketplaces or interstate commerce. So, there are a lot of challenges with a state-by-state approach. And those are the reasons that we have been so vocal about the need for a federal regulator and federal engagement – federal agencies that have consumer safety at the forefront to be engaged. I also think they have resources that states really desperately need. So every time there's a new cannabinoid product that comes out of hemp, then the list is growing. States don't necessarily have the scientific resources to look at the constituency of that product and to determine the safety of that product or what even is in that product. So those are areas where agencies like FDA generally are staffed with scientists that have the ability and the resources to do that. And we'd like to see a lot more engagement to support states with some of that science as well as regulation.

Andrew Peterson:
So then who should be the authority on *Cannabis* in the United States or the world? Should it be the DEA, the World Health Organization, or somebody more like the FDA?

Gillian Schauer (CANNRA):

Yeah, I think we've seen state models where the regulator is any number of agencies. We've got state programs that are regulated through public health, through revenue, through consumer protection, through public safety. So, states have really modeled that there can be a lot of different agencies that are involved in regulation. I think on the federal landscape, it would be challenging to have, say, the DEA be the sole regulator. There's a . . . very important social justice history to this particular substance that states have taken to heart. And I think it's really important that we do the same federally. And you know, the war on drugs did not impact everybody equally. And I think that it probably would not be the wisest decision to have the DEA be the regulator and to have it be so focused on enforcement. We spent a lot of hours at CANNRA talking about compliance and enforcement. We've seen a lot of states shift their compliance and enforcement approaches to be more education-based. With that said, you do really still have to have some tools for enforcement for the most egregious violations or drug trafficking. There's a role for that, but I don't think that should be the predominant role. We've seen a number of public health agencies involved in regulation. I think, however, public health agencies often don't have regulatory experience. They're typically not regulating a lot of other things. So, the most natural place for this to go might be somewhere like FDA, where they have a public health and consumer safety focus, but they also are regulators of other substances. The challenge is this is a very fast-paced area, as we've talked about, and we do not have any luxury of waiting a decade for regulation to happen. Regulation was needed yesterday, especially for things like hemp. So, you know, if it is an agency like FDA, I think there would be a need to work through a timeline that addressed the needs of today. There would also be a need to address a regulatory process that isn't rooted in having all the science. So traditionally, I think federal regulation has waited until we have a complete body of science to make any regulation. And states have learned that we don't have that luxury. You have to create regulation to protect consumer safety, without all the science. We're missing science in so many areas. If any students are listening and want to go into research, I think it'll be a very ripe career for them. And regulation has to go forward before we have the compendium of research to support different approaches. And regulation has to be nimble to recognize when we get new science. Hey, we may have gotten this wrong, we may need to adjust something. And I think that's the exact reason why you see even in states like Washington and Colorado that are 10 years in to legalizing adult-use marijuana, you see 15, 20 or more bills every legislative session in the state because they're making course corrections. They're adjusting for things that we've learned for new products that have come out. So, there's a need to have whoever regulates this federally be an agency that can be nimble, that can work with the science we have today, and adjust for science that we may get in the future.

Andrew Peterson:
Thanks.

Jahan Marcu:

Wow. I definitely can see a lot of challenges facing regulators with, you know, *"this agency's in charge", "No, this agency's in charge", "This is the one that should be."* Where's the policy? I've got to imagine this translates to a number of challenges for the operators who, as you pointed out, right? We have, it's like you have THC's as of right now, it's found in schedule one, schedule two, schedule three. It's found unscheduled in hemp. It's found in medical *Cannabis* and adult-use products. You know, we have the similar chemicals, similar active ingredients, if you will, and yet, what do you think some of the challenges or issues that *Cannabis* operators should be focusing on or what's ahead of them for the next 12 months or so?

Gillian Schauer (CANNRA):
Yeah, and we spend a lot of time at CANNRA focused on national stakeholders, including industry associations. So, I do spend a lot of time listening to challenges on the industry side. I think that's a really important part of any regulatory agency. I think the hemp reality has really been challenging to, you know, we've got a bunch of operators that are trying to play by the rules, which for marijuana, I'm going to use marijuana because it's to distinguish it from hemp, they're both *Cannabis*. We have a bunch of operators on the marijuana side that have invested a lot of money to comply with the regulations. And the regulations have been set to try to protect consumer safety. Marijuana products are also taxed on the adult-use side and the taxes can make those products more expensive. So, it can be extremely challenging to have hemp out there where there's no tax. There are no regulations to comply with. You don't have to pay a bunch of money or be in a lottery to get a license. So, we have marijuana operators that have been trying to play by the rules that now are watching folks come in on the hemp side and make the same products with a very different set of rules. So, I think there have been a number of financial challenges for the industry, in part due to hemp, but also just due to the nascent nature of this industry. And regulators are well aware of that, and I think are constantly trying to figure out how they can support their markets while still upholding consumer safety and public health goals. So, I think that's one of the biggest challenges to the industry – in just the financial aspects of trying to play in a new space when there are some competitors out there playing by different rules. I think there've also been challenges in trying to create an equitable marketplace. And that's something that we talk a lot at CANNRA about. As I said earlier, the war on drugs did not impact all communities equally. And it's, I think, extremely challenging to think that there are still people who are locked up for *Cannabis*-related offenses when others are making millions of dollars in the *Cannabis* industry. So we need to remedy that. And I think part of that is creating equity in the marketplace as well so that the people who have a chance to make money from selling *Cannabis* are not, you know, for lack of a less-refined term, not just a bunch of white men, but are representative of the communities that have been disproportionately impacted. And states are still learning a lot about how to do that and how to do that well, and how to do that within the confines of the law. So that's an area that I think is a challenge for the industry. And then I think, you know, the industry, as I said earlier, will be challenged by the federal uncertainty. And it I think state regulators will be doing everything they can to try to have certainty in their state markets, where they can. But it's uncertain what will happen with interstate commerce. It's uncertain what will happen with potential rescheduling that could, as many argue, put *Cannabis* on more of a pharmaceutical pathway. And what does that mean for state markets? And will state markets just continue to operate as they are? Or will there be a slow funneling into a different supply chain for *Cannabis* that would be really detrimental to a lot of the operators now. So it's not an easy space to be an operator in this industry. And I think regulators hear and understand that. And I

haven't met a regulator who isn't interested in learning from the folks that are licensees in their state. I think that's an important way to continue to move forward in policy.

Jahan Marcu:
And I just wanted to ask just a quick follow up. Interstate commerce, do you think that will happen with or without – that it will move forward in terms of policy with or without rescheduling of *Cannabis*? Do you think that that's just going to continue or is it dependent on rescheduling?

Gillian Schauer (CANNRA):
I don't think it's dependent on rescheduling. And I actually think personally – moving from like a schedule one to a schedule three, I don't know that really does a whole lot to impact interstate commerce. But I think there are a lot of irons in the fire now that suggest that interstate commerce might come sooner than we think. So, we've got a bunch of court cases challenging on the dormant commerce clause. That is one area where scheduling might change the arguments made in court. But basically, court cases saying . . . X, Y, Z state, you can't have a program that's just in your state because that violates the dormant commerce clause. And we've seen some decisions that suggest that even though marijuana is still schedule one and that dormant commerce clause does apply in some instances. So we could see some court cases and a judicial pathway through which we get to interstate commerce. We also see a number of states now, mostly supply states that have a lot of growing that's happening, that have passed legislatively different approaches to try to have the state be able to opt into interstate agreements where, say, the state of California, which is a prominent state that has this policy in place, might talk to some southern state that's just setting up a market and say, hey, we will be your grower, we will be your supplier, let's set up some sort of contract to do that. So we could see that avenue move forward as well. And then I think we continue to have the congressional avenue, where we could see de-scheduling by Congress, which would open up interstate commerce. We also could see something like another coal memo that says, we're not going to challenge interstate agreements that might come about. And I think finally, we already have interstate commerce, if we're being honest. We have it through hemp.

Jahan Marcu:
[Laughs]

Gillian Schauer (CANNRA):
– then we have it through the illicit *Cannabis* market. It's just not within a regulatory framework at present. So I think we have a lot of places to look to see where we would need to overlay appropriate regulation to protect consumers just in the market that we have today.

Andrew Peterson:
Dr. Schauer, you've got a lot of things going on, and I appreciate all the balls that you have up in the air, and many of them probably keep you up at night.

Gillian Schauer (CANNRA):
Yes, they do.

Andrew Peterson:
But if you woke up one day and said, oh good, that will make my job easier, what would that be?

Gillian Schauer (CANNRA):
I think there are two things. The first one is a greater willingness from anyone in this policy space to listen and learn from others. And we have tried to facilitate that. We have, at CANNRA, external stakeholder meeting that, on purpose, we invite a bunch of national stakeholders that disagree with each other, that may not often be in the same room. And we invite them to share their positions with our members so that our members can hear all different positions and the context behind them. And so they can hear a dialogue between people who might disagree. And I think that's really important at the beginning of a new legal industry that we have those conversations happening across groups that disagree so that we can identify areas of overlap, places where people might actually agree and they didn't even realize it. So I think more cross-cutting diverse discussions would certainly make this space easier. And we're trying to facilitate those at CANNRA, and we facilitate those among regulators too. I mean, we have red states, blue states, purple states; we have states with different political leanings, different policies, different . . .

Gillian Schauer (CANNRA):
. . . populations. And we're trying to convene regulators to learn from each other in those different realities. It would be great to see that happen in a broader scale across folks engaged in the *Cannabis* policy and regulatory space. So that would be the first one. I think the second one would be federal engagement. We work regularly with a lot of different federal agencies, but they are extremely limited in their ability to work with us. And I would love to see that change so that we can have federal agencies, again, with many greater resources, engaged in helping our members on the science side, engaged in helping our members with different things, even basic questions like what pesticides should we be using on *Cannabis*? What are the implications of using different pesticides? That's an area where EPA would normally be engaged and we'd love to have more of that engagement. Data monitoring and data collection to understand the implication of policies – that's an area where CDC . . . is heavily engaged across so many other domains, and we'd love to have them be able to be much more engaged. As I mentioned earlier, FDA and their lab and analytical ability, that's desperately needed for states. So having federal agencies be sort of released and allowed to

work more with states on this topic would be huge. And then I think the third and final one is research, And . . .

Gillian Schauer (CANNRA):
. . . figuring out what we can do to fast track research and make research reflective of real world situations. And we do work, you know, closely with NIH agencies and I know they want to do that as well. But a lot of the research that's happened is not reflective of the product landscape that we see in the field, the reality of how consumers are accessing and using the products, and we really need to catch up because that will help inform science-based policy, which as a scientist myself, I believe is really important.

Andrew Peterson:
Thank you.

Jahan Marcu:
All right. Well, in closing, the final question for you is a resource question. Is there a resource you find yourself most often sharing with people, you know, like a book, a podcast, article, something else, a YouTube video, an infographic, um, you know, cause you mentioned a lot about communicating and listening and I'm just wondering, do you have a go-to thing you send people or resource?

Gillian Schauer (CANNRA):
Yeah, you're right. I am very often on the listen-and-receive mode. So that's a tough question for me to think about what I send out. And selfishly, I would probably say the thing I send out the most is our website because we're still a new association and we've got a lot of letters we've written to Congress and resources, you know, all of the summer of hemp activities, our written testimony, our 42-page RFI, those are all on our website.

Jahan Marcu:
Wow, that's fantastic!

Gillian Schauer (CANNRA):

So it's www.cann-ra.org And I hope that it will become a place that more people will know about and go to. Again, we are a newer association, so it's logical that it would take some time, but we're trying to share a lot, and we're trying to share things that are, again, fact-based and from a nonpartisan position, and I think there's a need for that.

Andrew Peterson:
Well, thank you so much for sharing all of this information today about CANNRA, about all the work that you and the organization is doing, and plus your insights into the *Can-*

nabis and the hemp industry. I wish you the best in all that you're doing because I know what you're doing is only going to move this industry forward. So I really do appreciate that. And thank you for that.

Gillian Schauer (CANNRA):
Thank you both for having me, it's an honor.

Thomas A. Trite, Tzuo-Zen Lee, Andrew M. Peterson

Will *Cannabis* remain a standalone industry?

Introduction

In this chapter, we will examine whether *Cannabis* will remain a standalone industry, which is dependent on several legislative and regulatory issues that will be highlighted. Other factors that affect the industry's independence will be addressed, as well as all relevant aspects that either add to or subtract from that independence.

After reviewing the information provided here, the reader will have the necessary information to formulate their own conclusion. Only time will determine the result, but either outcome will provide interesting opportunities in a regulated environment.

Federal and State Regulation of *Cannabis*

First, we need to explore the federal and state status of medical and/or adult-use *Cannabis*. Figure 1 shows each state's *Cannabis* approval in 2022. Federally, *Cannabis* is currently classified as a Schedule I drug and, therefore, at this point, is illegal. As of April 24, 2023, there are 38 states, three territories, and the District of Columbia that permit the medical use of *Cannabis*. Additionally, as of June 1, 2023, there are 23 states, two territories, and the District of Columbia that have enacted measures to regulate *Cannabis* for adult use, also known as recreational *Cannabis*. There are nine states that allow the use of "low-THC, high-cannabidiol (CBD)" products for medical reasons in limited situations or as a legal defense [1, 2].

Cannabis industry in comparison

In order to address whether *Cannabis* will remain a standalone industry, let us first take a brief look at the current industry in comparison to other industries. This section delves into an intriguing comparative analysis of several industries, such as cigarettes, over-the-counter (OTC) medicines, coffee, salty snacks, and others. By exploring sales figures, employment statistics, and growth trajectories across these industries, we can gain valuable insights into the *Cannabis* industry's standing and potential as a standalone economic powerhouse. While each industry brings its distinctive attrib-

Thomas A. Trite, Tzuo-Zen Lee, PA Options for Wellness Clinical Registrant
Andrew M. Peterson, Philadelphia College of Pharmacy, Saint Joseph's University, Philadelphia, PA

https://doi.org/10.1515/9783110677508-004

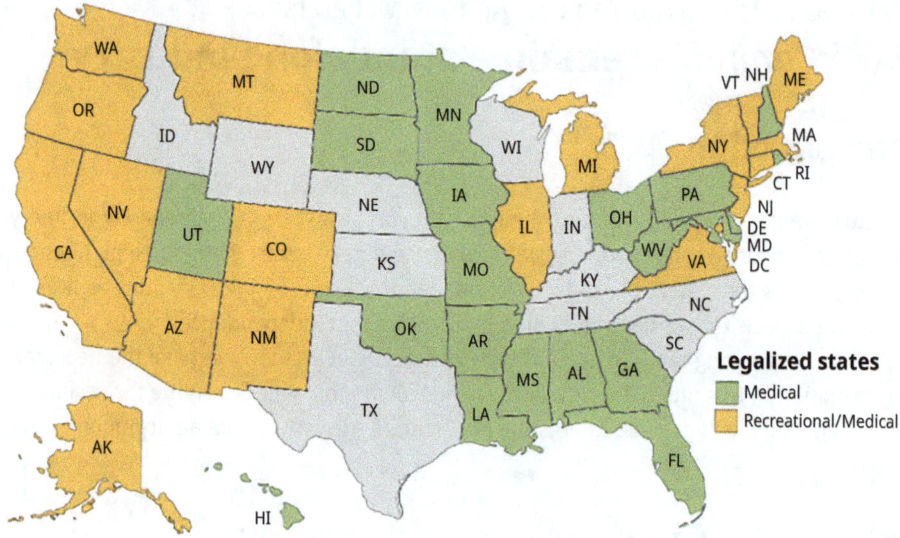

Figure 1: Breakdown of States with Medical and Recreational State Approvals.
Note: Does not include states that have legalized only CBD-based oils.

utes and challenges, understanding their relative positions and trajectories in the market is crucial for comprehending the *Cannabis* industry's place in the broader economic landscape of the United States.

Sales

For a clearer perspective of total sales, please see Figure 2 which shows that medical, as well as adult-use *Cannabis* sales are the third largest within those industries surveyed, behind cigarette and OTC1.1 medicine sales [3]. It is interesting to note that *Cannabis* sales are larger than Starbucks North American sales by more than $4B. Figure 3 shows the significant projected growth in *Cannabis* sales from 2021 through 2026. While there is a moderate but steady growth in the medical aspect (8.2%), the compounded annual growth rate (CAGR) in the adult-use aspect of the business is predicted to be double that of the medical side at 17.5%.

Employment

If we look at employment opportunities for 2021, as noted in Figure 4, the *Cannabis* industry was the fourth largest supplier of jobs within surveyed industries, falling be-

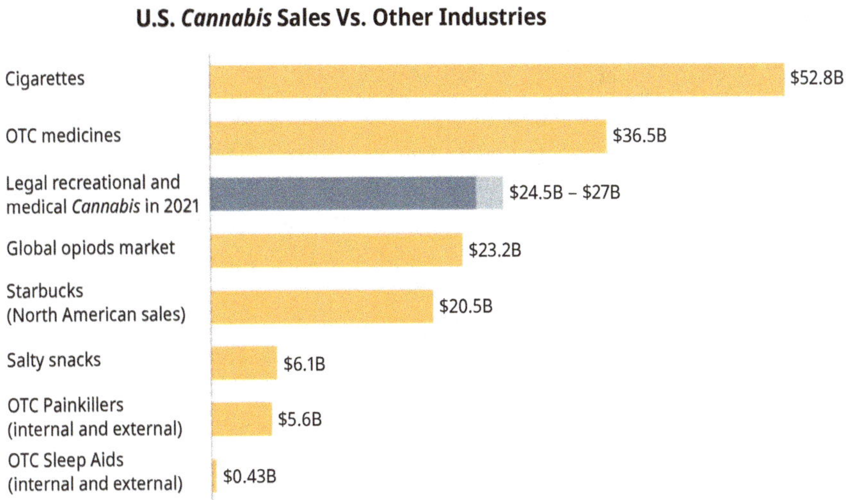

Figure 2: Listing of Several Industries and Leading Companies for Perspective of Total Sales.
NOTE: All data reported for 2021. Source: MJBizFactbook research, Statista, Consumer Healthcare Products Association, BCC Research, Mordor Research, Allied Research, Starbucks financial filings, IRI Market Advantage.

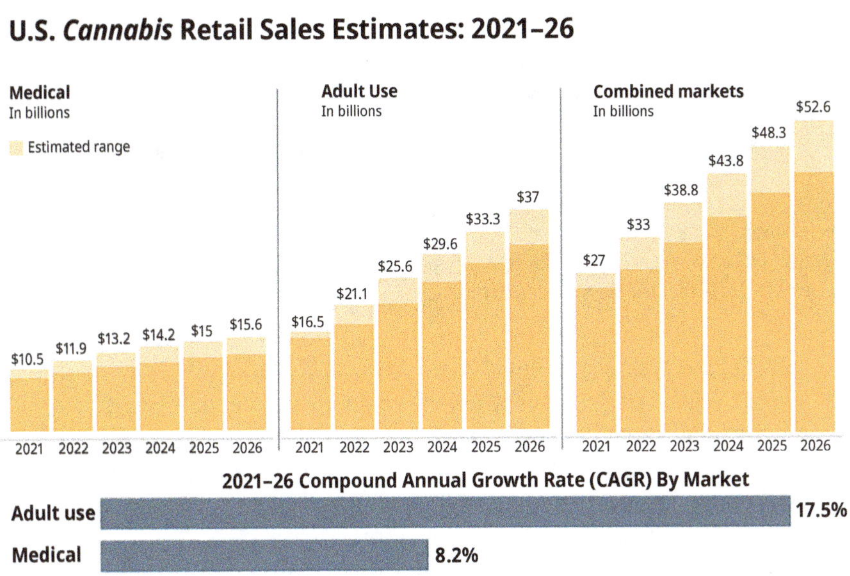

Figure 3: U.S. *Cannabis* Retail Sales Growth Estimate: 2021–2026.
Source: MJBizFactbook.

hind elementary school teachers, lawyers, and bartenders [4]. The number of *Cannabis* employees is now larger than real estate sales agents, dishwashers, counselors, and Starbucks providing 360,000 to 440,000 jobs and growing. This not only adds to the economy through sales, but also to the overall economy with income from its staff fueling the local, state, and ultimately, the national economy. This shows that *Cannabis* is a growing industry, while many others are facing declines, with the number of *Cannabis* jobs continuing to increase, being largely sought after in this extremely competitive market.

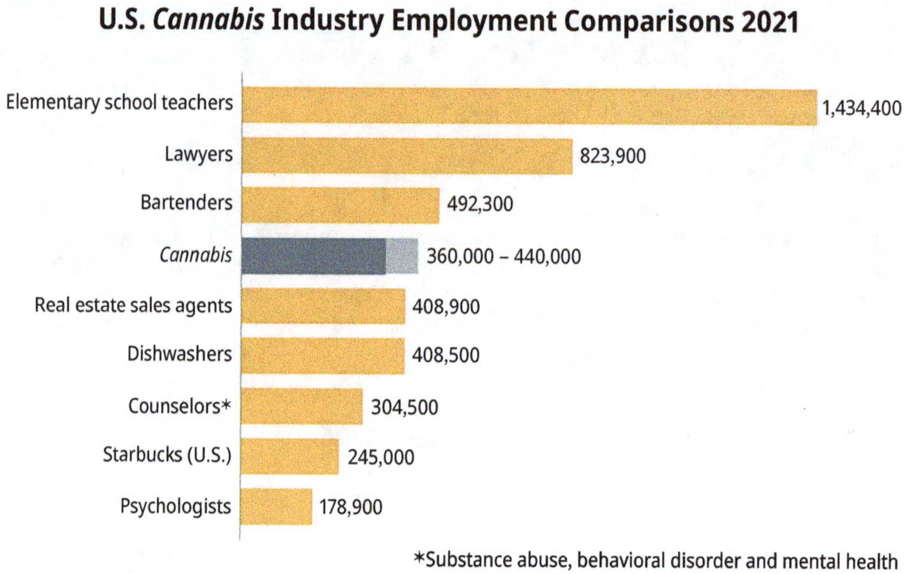

Figure 4: Analysis of Industry Professionals.
Source: Pew Research Center.

Additionally, Table 1 provides recent statistics from similar industries that may incorporate *Cannabis* into their products while also showing those that might possibly become the most competitive [5]. The *Cannabis* industry's growth can be attributed, in large part, to the integration of *Cannabis*-infused products by neighboring sectors such as food, snacks, alcohol, pharmaceuticals, and tobacco. This trend emerges from shifting consumer preferences and evolving regulations that present an opportunity for synergistic growth. For instance, multinational alcohol and beverage giants are exploring *Cannabis*-infused beverages, a move that could open doors to new distribution channels and amplify the industry's reach. Similarly, pharmaceutical conglomerates are venturing into cannabinoid-based medications, creating possibilities for cutting-edge innovations backed by their robust research capabilities. This positive interplay

Table 1: Compilation of Various Related and Potentially Competitive Major Industries in the United States.

Industry	Revenue (billions, $)	Profit (%)	Employees (thousands)	Wages (billions, $)
Pharmaceuticals				
Brand	239.0	7.9	278.0	29.50
Pharmaceutical wholesale	1400.0	1.2	331.0	53.70
Pharmacy	569.7	3.9	1000.0	54.00
Compounding pharmacy	10.0	4.1	20.0	0.90
Beverage				
Wine/spirit wholesale	130.4	3.4	101.0	9.20
Beer/wine/liquor mfg.	78.9	3.8	190.0	5.48
Coffee/snack shops	54.5	3.4	922.0	16.50
	Starbucks (38.3% of coffee/snack Industry)			
Food				
Candy	12.9	4.0	27.3	1.80
Tobacco	52.8	39.8	10.3	0.97
Cannabis				
CBD	2.4	−5.0	24.7	0.73
Med/rec *Cannabis*	22.6	7.0	219.0	5.60
Industrial hemp	0.9	7.8	36.0	0.47

mirrors strategies seen in industries like tobacco and food, which have adapted over the years to changes in consumer preferences and tastes. As these sectors align with *Cannabis*, the *Cannabis* industry stands to gain from shared expertise, expanded market access, and collective innovation.

Cannabis: Not just a single business

Businesses within the *Cannabis* industry fall primarily into two subunits: plant-touching and ancillary services as seen in Figure 5. Plant-touching businesses, simply put, are those that interact directly with the plant. This includes dispensaries, cultivators, manufacturers, processors, and vertically integrated businesses. In the next few sections, we will look at a breakdown of the various operating units and processes that make up the *Cannabis* business and the industry focus:

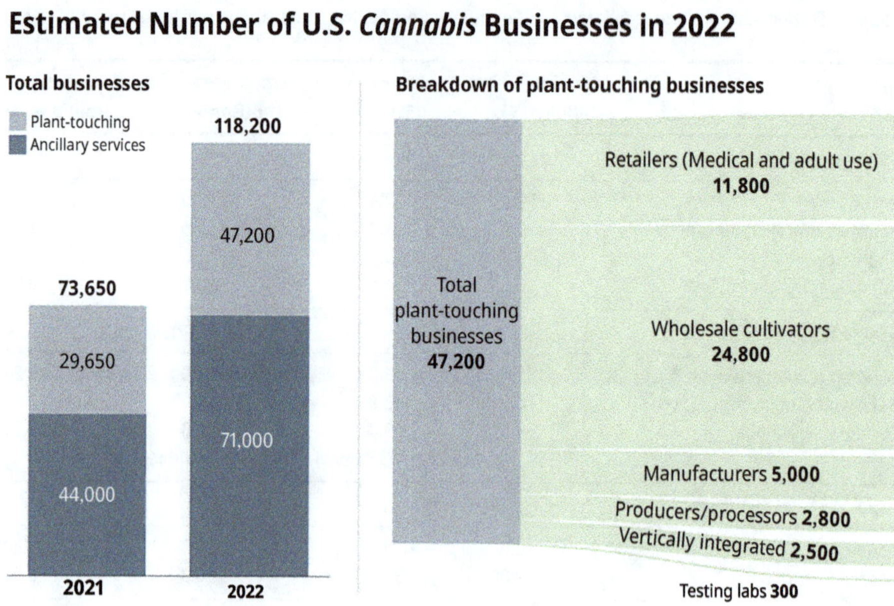

Figure 5: Estimated U.S. *Cannabis* Businesses in 2022.
NOTE: Estimates are high end of scale. Source: MJBiz Factbook.

Seeds and clones

The seeds and clones aspect of the *Cannabis* business is one of the critical operating units within the industry's system. This component deals with the provision of seeds or clones, which are essential for initiating the growth of *Cannabis* plants and for propagating future plants of every strain. The cultivation of these genetic foundations forms the backbone of the entire *Cannabis* production process.

Moreover, the development of cultivars, with specific genetic make-ups that yield higher quantities of select cannabinoid components, presents a remarkable opportunity for advancing medical treatments. By focusing on producing strains with enhanced levels of therapeutic cannabinoids, such as cannabidiol (CBD) or tetrahydrocannabinol (THC), the *Cannabis* industry can contribute significantly to the treatment of various medical conditions. These potent cultivars could potentially offer more effective and targeted relief for patients suffering from ailments like chronic pain, epilepsy, anxiety, and other medical issues where cannabinoids have shown promising therapeutic effects.

As the *Cannabis* business expands and gains acceptance in the medical community, the seeds/clones unit plays a vital role in not only meeting the demands of the market but also in furthering scientific research and advancements in the field of medical *Can-*

nabis. Its capacity to create tailored genetic profiles that address specific medical needs underscores its indispensable role within the broader *Cannabis* ecosystem. As a result, focusing on the business aspect of this component involves investing in research and development to optimize and refine genetic profiles, partnering with medical professionals and researchers, and ensuring compliance with regulatory standards to provide safe and reliable cultivars for both medicinal and recreational uses.

Cultivation

The process of growing *Cannabis* plants in a safe and efficient manner impacts the overall quality of the product and, ultimately, improves patient outcomes. By focusing on achieving excellence in cultivation, the *Cannabis* industry can ensure the production of high-quality, consistent, and reliable *Cannabis* products that meet the needs of consumers and patients alike.

Efficiency plays a pivotal role in this aspect, and it can be attained through a strategic integration of the best current cultivation practices with advanced automated processes and building management systems. These innovative techniques enable cultivators to create and maintain optimal conditions for the *Cannabis* plant at every phase of its growth cycle.

By leveraging cutting-edge technologies, cultivators can precisely monitor and control crucial environmental factors such as temperature, humidity, light exposure, nutrient delivery, and irrigation. This level of precision ensures that the plants receive the ideal conditions needed for their healthy development and maximum cannabinoid and terpene production. Moreover, the implementation of automated systems reduces the risk of human errors and ensures consistency in the cultivation process. This consistency translates into a more predictable and standardized end product, which is essential for both medical patients and recreational users seeking reliable and reproducible effects.

Creating a safe and controlled cultivation environment also helps minimize the risk of contamination, pests, and diseases that could jeopardize the crop. It enables cultivators to adhere to strict quality control measures, ensuring that the final *Cannabis* products are free from harmful substances and contaminants, thus enhancing the safety and well-being of patients.

Furthermore, when cultivators prioritize safe and efficient cultivation practices, it fosters continuous improvement and innovation in the industry. Research and development efforts can focus on finding even better methods to optimize plant growth, enhance cannabinoid profiles, and improve the overall therapeutic potential of *Cannabis*.

Extraction and processing

The extraction and processing aspect of the *Cannabis* business is a pivotal step that holds significant potential for identifying and harnessing the full spectrum of cannabinoids, terpenes, and flavonoids present in the plant. This process not only enables the extraction of various beneficial compounds but also plays a crucial role in formulating products specifically designed to enhance patient outcomes and ensure safety.

Through sophisticated extraction techniques, *Cannabis* manufacturers can isolate and concentrate specific cannabinoids like CBD, THC, cannabigerol (CBG4.4.2), and others. Additionally, the extraction process allows for the preservation of valuable terpenes and flavonoids, which contribute to the entourage effect – the synergistic interaction between these compounds that enhances the overall therapeutic potential of *Cannabis*-based products.

By producing extracts with precise compositions, *Cannabis* companies can tailor their offerings to target specific medical conditions and patient needs. For instance, patients seeking relief from chronic pain may benefit from products with higher CBD-to-THC ratios, while those dealing with insomnia or anxiety might find better results with different cannabinoid profiles. This customization empowers medical professionals and patients to select products that align with individual treatment plans, thus optimizing the potential for positive health outcomes.

Moreover, the extraction and processing phase is essential for ensuring product safety. By using advanced technology and adhering to rigorous quality standards, manufacturers can produce purified extracts that are free from harmful contaminants and impurities. This commitment to safety is particularly critical for medical *Cannabis* patients with compromised immune systems or other health concerns.

Furthermore, this aspect of the *Cannabis* business fosters continuous research and development efforts. Scientists and researchers can explore various extraction methods and processing techniques to improve the efficiency of cannabinoid isolation and to discover novel compounds with potential therapeutic benefits. This ongoing exploration is vital for expanding our understanding of the *Cannabis* plant and its medical applications.

Manufacturing

The manufacturing component of the *Cannabis* business encompasses all stages of product production, with a primary focus on ensuring the safety and proper dosage forms to deliver products in the preferred method of administration for consumers and patients. Integrating the best manufacturing technologies from traditional industries into the *Cannabis* product manufacturing processes is of paramount importance. By doing so, manufacturers can achieve higher levels of consistency in purity and composition throughout their product lines. This technological integration helps in maintaining standardized production methods, which is essential for creating reliable and predictable products.

The emphasis on consistency in quality and purity holds immense significance, especially in the medical *Cannabis* field. When medical professionals and patients can rely on consistent formulations and standardized dosages, it facilitates a more targeted approach to treating medical conditions with *Cannabis*. Specific conditions may require precise ratios of cannabinoids and other compounds, and having products with consistent purity ensures that patients receive the same therapeutic effects with each use.

Moreover, the focus on manufacturing excellence contributes to minimizing side effects. When products are formulated with precise and consistent compositions, patients can have better control over their treatment, reducing the risk of adverse reactions due to inconsistent dosing or impurities.

By embracing advanced manufacturing technologies, *Cannabis* companies can adhere to strict quality control measures, ensuring that products are free from contaminants and impurities that could compromise their safety and efficacy. This dedication to producing safe and reliable products enhances consumer confidence and strengthens the position of *Cannabis* as a legitimate medical option.

Furthermore, as the *Cannabis* industry continues to evolve, the integration of cutting-edge manufacturing technologies drives innovation and efficiency. Researchers and manufacturers can collaborate to explore novel methods of production, novel delivery systems, and new dosage forms to expand the range of *Cannabis*-based treatment options available to patients.

Packaging

The packaging aspect of the *Cannabis* business plays a critical role in ensuring the safety and proper identification of products, making it a crucial element of the overall *Cannabis* industry. By implementing effective packaging and labeling practices, *Cannabis* companies can instill trust and confidence in consumers and patients alike.

Proper packaging is essential to protect the integrity of *Cannabis* products during storage, transportation, and handling. It helps maintain the freshness and potency of the products, ensuring that consumers receive *Cannabis* in its optimal condition. Moreover, well-designed packaging prevents contamination and tampering, guaranteeing that consumers receive safe and unadulterated products.

In addition to safety, accurate labeling is paramount in the *Cannabis* industry. Labels must provide essential information about the product, including its potency, dosage, instructions, potential allergens, and other relevant details. This information empowers consumers to make informed choices about their *Cannabis* use and helps medical professionals recommend appropriate products to patients based on their individual needs.

Integrating current pharmaceutical technologies into *Cannabis* packaging and labeling further enhances patient safety. By adopting technologies commonly used in the pharmaceutical industry, *Cannabis* companies can implement tamper-evident seals, child-resistant closures, and other features that enhance product security. These measures protect against unauthorized access to the product and reduce the risk of accidental ingestion by children.

By leveraging pharmaceutical packaging technologies, *Cannabis* companies can implement serialization and track-and-trace systems. This enables accurate product identification and tracking throughout the supply chain, ensuring that consumers and patients receive authentic and regulated *Cannabis* products. Such measures are particularly important in areas where *Cannabis* is legal for medical use, as they help prevent counterfeiting and ensure patients have access to safe and legitimate treatments.

The integration of pharmaceutical technologies also supports regulatory compliance, as it enables *Cannabis* companies to adhere to strict packaging and labeling requirements set forth by governing authorities. Compliance is essential for maintaining the reputation of the *Cannabis* industry and securing its position as a legitimate player in the medical and recreational markets.

Transportation

The transportation aspect of the *Cannabis* business supply chain is responsible for efficiently delivering *Cannabis* products to dispensaries and, in certain states, directly to patients. The timely and reliable transportation of these products is essential to meet consumer demand and ensure that patients have access to their prescribed treatments when needed.

Integrating current logistics tools and practices is paramount to streamline and optimize the delivery process. By adopting advanced transportation technologies, such as GPS tracking, route optimization, and real-time monitoring, *Cannabis* companies can enhance the efficiency and accuracy of their deliveries. These tools help in planning the most efficient routes, minimizing delivery times, and reducing transportation costs.

Moreover, the integration of logistics practices used in other industries enables *Cannabis* companies to benefit from established best practices in the transportation sector. Lessons learned from traditional delivery systems can be adapted to suit the unique needs and regulations of the *Cannabis* industry, ensuring that products are transported safely and securely. Integrating logistics tools that provide real-time updates and data can aid in ensuring compliance with legal requirements, such as tracking the movement of products and adhering to licensing and documentation procedures.

Efficient transportation also plays a significant role in managing inventory and preventing stockouts. Timely deliveries to dispensaries and patients help maintain a consistent supply of *Cannabis* products, reducing the risk of product shortages and keeping customers satisfied.

In states where direct-to-patient delivery is permitted, efficient transportation becomes even more critical. Patients relying on medical *Cannabis* for their treatments often have specific medical needs and may depend on regular deliveries to manage their conditions effectively. Implementing reliable transportation practices is crucial in ensuring that patients receive their medication on time, improving their overall treatment experience.

Dispensaries and retail sales

The retail sector is a pivotal component of the *Cannabis* industry, encompassing the training and education of patients, staff, and consumers on safety and the selection of the best products to achieve their desired outcomes. In the medical context, dispensaries play a critical role in improving patient outcomes and enhancing their quality of life. Through well-trained staff, patients can receive valuable guidance on selecting *Cannabis* products that align with their specific medical needs. Dispensaries offer a range of products with varying cannabinoid profiles and delivery methods, allowing medical professionals and patients to tailor treatments to address specific conditions or symptoms. Additionally, staff can provide dosage recommendations and educate patients on potential interactions with other medications, ensuring safe and effective *Cannabis* use in conjunction with existing medical treatments. This personalized approach fosters a more targeted and individualized treatment experience for patients, leading to potential improvements in their overall well-being.

For adult-use consumers, dispensaries also offer a safe and controlled environment for the responsible use of *Cannabis* for social and recreational purposes. As with the medical side, trained staff can provide valuable information about different *Cannabis* products, their potency, and their potential effects, allowing consumers to make informed choices based on their preferences. Dispensaries promote responsible consumption, encouraging customers to be aware of dosage and potential side effects to ensure a positive and enjoyable experience. Moreover, dispensaries can offer a diverse range of products to cater to various consumer preferences, such as edibles, vapes, tinctures, and more, allowing for personalized and enjoyable *Cannabis* use.

The training and education provided by dispensaries extend beyond consumers to their own staff as well. Well-informed dispensary staff can confidently assist customers, providing accurate information and guidance, thus elevating the overall dispensary experience for visitors. Proper staff training also ensures compliance with regulations, safety protocols, and ethical business practices, contributing to the industry's professionalism and reputation.

Research

The research aspect of the *Cannabis* industry is instrumental in advancing our understanding of this complex plant and unlocking its full potential for medical and commercial applications.

Medical research

In the medical domain, research focuses on identifying additional cannabinoids and terpenes that hold therapeutic promise in addressing various medical symptoms and specific diseases. By conducting proper scientific studies, researchers aim to provide conclusive evidence of the efficacy of *Cannabis* in treating specific conditions, moving beyond anecdotal claims to establish solid clinical data. This rigorous approach is essential for gaining acceptance within the medical community and supporting evidence-based treatment decisions.

Through medical research, scientists seek to explore the synergistic effects of cannabinoids and terpenes, known as the entourage effect, to understand how different combinations can yield enhanced therapeutic outcomes. Identifying these synergies can lead to the development of targeted *Cannabis*-based medications with potentially fewer side effects and improved efficacy compared to single compounds. Utilization for evidence-based information in patient care is critical in providing optimized results for the patient. Information that is compiled from past research and updated from current findings should be easily accessible to health professionals and patients. Tools that facilitate this access will be invaluable.

Plant research

On the plant side of research, scientists strive to identify and characterize the various cannabinoids, terpenes, and flavonoids present in different *Cannabis* strains. Understanding the chemical profiles of these strains enables cultivators to produce products that cater to specific needs and preferences. By identifying strains that exhibit superior therapeutic potential, researchers can optimize cultivation practices to yield higher quantities of beneficial compounds, thus improving the overall quality and safety of *Cannabis* products.

Plant-focused research explores cost-effective methods to cultivate *Cannabis* and extract valuable compounds, which is crucial for a sustainable and economically viable production. As the industry grows, efficient cultivation and extraction techniques become increasingly essential to meet the rising demand for *Cannabis* products.

By fostering research in both medical and plant aspects, the *Cannabis* industry can push the boundaries of knowledge and innovation. This research not only supports the development of safe and effective medical treatments but also drives advancements in cultivation practices, product development, and regulatory standards.

Furthermore, research in these areas contributes to the overall destigmatization of *Cannabis* by providing scientific evidence to support its potential benefits and safety. As research findings become more widely accepted and acknowledged, *Cannabis* is likely to gain further recognition as a valuable and legitimate option for medical treatments.

Rescheduling *Cannabis*

As you can see, each of the above operating units comprising a *Cannabis* operation is a function that can be found within other industries. This begs the question – why are *Cannabis* companies treated differently? Simply put, *Cannabis* is considered a Schedule I drug.

Cannabis classification as a schedule I substance (A drug or other substance that has a high chance of being abused or causing addiction and that has no FDA-approved medical use in the United States.) has tremendous banking, funding, and tax consequences. Just as in the healthcare industry, specifically referencing pharmacy (as it is the best comparator), there will be many areas of expertise to pursue. The first areas to address are the legislative and regulatory next steps in order to move forward. See Figure 6 for a timeline of legislative reform for *Cannabis* [7]. As of the writing of this chapter, the United

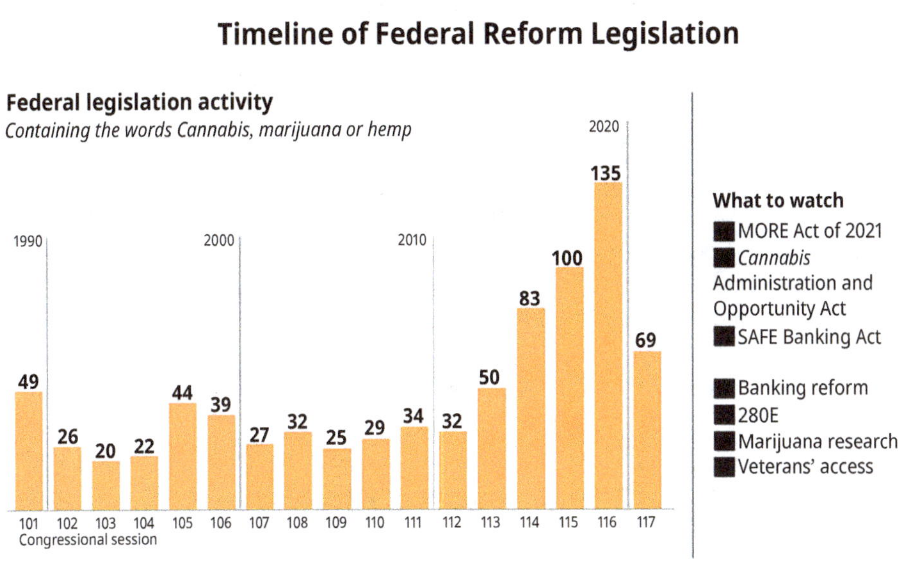

Figure 6: Federal Reform Legislation.
Source: MJBiz Factbook.

States Department of Health and Human Services, through the Food and Drug Administration (FDA), has recommended that *Cannabis* be rescheduled to a Schedule III drug.

On December 2, 2022, President Biden signed into law the Medical Marijuana and Cannabinoid Research Expansion Act, HR8454, also known as The *Cannabis* Research Bill, to remove obstacles in the research of *Cannabis*. This bipartisan legislation is the first standalone *Cannabis* reform bill to pass both the House and Senate, making a significant milestone in federal *Cannabis* policy.

But by far, the single most effective step that needs to be taken is to reschedule *Cannabis* from a Schedule I to a Schedule III drug, similar to that of Marinol (Dronabinol – synthetic THC), which produces the psychological effects of *Cannabis*. This step would allow for some improved access, better bottom-lines and even greater access to the drugs. This is akin to the reclassification of Epidiolex as a noncontrolled substance. Epidiolex (cannabidiol) is the first and only FDA-approved prescription of CBD to treat seizures associated with Lennox-Gastaut Syndrome (LGS), Dravet Syndrome, or Tuberous Sclerosis Complex (TSC) in patients one year old or older.

Finance and banking legislation

There are several key actions that would place *Cannabis* on a level-playing field, all dependent on it being removed federally from Schedule I status to Schedule III status. This move would impact key regulations, including the Safe Banking Act and IRS Tax Code 280E.

The Safe Banking Act is a bill aimed at providing a safe, secure way for banks and credit unions to conduct business with *Cannabis*-related companies in states where *Cannabis* is legal. By passing the Safe Banking Act, the *Cannabis* industry would be treated in the same manner as all other businesses, and it will allow more banks to do business with *Cannabis* clients as well as reduce the bank charges incurred by *Cannabis* companies, making them comparable to other industries.

IRS Tax Code 280E speaks to expenditures that are associated with the "illegal sale of drugs" and applies to *Cannabis* business, so long as *Cannabis* remains a Schedule I drug. It prohibits *Cannabis* businesses from deducting otherwise ordinary business expenses from gross income associated with the "trafficking" of Schedule I or II substances.

Thus, *Cannabis* business expenses such as payroll, benefits, leases, utilities, supplies, and dispensary operating expenses are not able to be deducted. By reclassifying *Cannabis* to Schedule III, these expenses would become deductible, as permitted for other industries. If they only move it to Schedule II, it would not resolve 280E and the deductions for the above that are afforded to every other business would still not be allowed, and IRS taxes would remain a huge disadvantage compared to all other businesses.

These issues, as well as several legislative and executive actions on the horizon, will determine not only how independent the *Cannabis* industry will remain, but also how successful would be the industry as a whole.

Current and future opportunities for innovation

There are numerous areas for professionals to get involved in the *Cannabis* business. Table 2 is a listing of those areas that will be further discussed in the following section.

Table 2: Current and Future Opportunities for Innovation.

- Consulting
  - Patients
  - Industry
  - Health Systems
  - Universities
  - Research
  - Compliance
- Dispensing
  - Pharmacy
  - Dispensary – medical
  - Recreational dispensary
- Manufacturing
  - Compounding
  - Personalized medicine
  - Development of meds and forms of administration
  - Other industries
    - Liquor
    - Beverage
    - Edibles
    - Tobacco
    - Cosmetics and skin care
  - Pharmaceuticals
- Research
  - Outcomes
  - New cannabinoids, terpenes, and flavonoids
  - Consumer preferences
  - Forms of administration

Consulting

Consultants can assist *Cannabis* businesses in understanding and adhering to the complex web of regulations and licensing requirements that govern the industry. This may involve ensuring compliance with local, state, and federal laws; obtaining permits and licenses; and implementing standard operating procedures. Consultants with expertise in horticulture, agronomy, and botany can provide guidance on optimal cultivation practices, including lighting systems., nutrient management, pest control, strain selection, and harvesting techniques, Extraction consultants specialize in the techniques and equipment used to extract cannabinoids, terpenes, and other compounds from the *Cannabis* plants. They can advise on methods such as solvent-based extraction, CO2 extraction, distillation, and other manufacturing processes. Consultants can also assist with the development and formulation of *Cannabis*-infused products, including topicals, concentrates, tinctures, vape cartridges, and edibles. They can provide expertise in product formulation, dosage consistency, quality control, and packaging design. Consultants with knowledge of retail operations can provide guidance on dispensary layout and design, inventory management, customer service, patient care, staff training, as well as marketing and compliance.

Consultants can also help businesses build and promote their brands in a highly regulated and competitive market. They can develop marketing strategies, create advertising campaigns, manage social media presence, and navigate restrictions imposed by regulations.

With the *Cannabis* industry's unique financial challenges, consultants specializing in *Cannabis* finance can assist businesses with financial planning, accounting, tax compliance, investment strategies, and fundraising. Consultants with scientific backgrounds can provide expertise in *Cannabis* research and development. They can assist with clinical trials, product testing, laboratory setup, and scientific documentation for regulatory purposes.

Dispensing

Promoting and increasing patient education is crucial in the medical *Cannabis* industry. Many patients may be new to *Cannabis* or unfamiliar with its potential benefits and usage. Dispensaries should prioritize educating patients about different strains, consumption methods, dosage guidelines, potential side effects, and interactions with other medications. Clear and accessible educational materials, one-on-one consultations, and workshops can help patients make informed decisions about their treatment.

Clear and comprehensive product labeling is essential to ensure patients have accurate information about the products they purchase. Labels should include cannabinoid content, terpene profiles, recommended dosages, and potential allergens. Standardized labeling requirements across jurisdictions can help patients compare products easily and make informed choices.

Dispensary staff may assist patients, answer questions, and provide guidance. Investing in ongoing training programs for staff can enhance their knowledge of *Cannabis* therapeutics, strain profiles, product options, and emerging research. Well-informed and trained staff can provide personalized recommendations and support to patients, enhancing their overall experience.

Building stronger connections between dispensaries and healthcare providers can also improve patient care. Collaboration with healthcare professionals, including physicians, nurses, and pharmacists, can facilitate knowledge exchange, promote evidence-based recommendations, and ensure coordinated care. Establishing communication channels and referral networks will enhance patient outcomes and promote a holistic approach to treatment.

Manufacturing

Standardized manufacturing processes ensure uniformity across batches and products. These processes encompass defining and adhering to specific procedures for cultivation, extraction, formulation, and packaging. Through meticulous documentation and strict adherence to standard operating procedures (SOPs), manufacturers can minimize variations and guarantee that each product meets the desired specifications.

Maintaining rigorous quality control measures is imperative throughout the manufacturing process. This involves regular testing for potency, purity, and potential contaminants such as pesticides, heavy metals, and microbial pathogens. Manufacturers may also assess the consistency of cannabinoid and terpene profiles. To uphold quality standards, establishments may setup in-house testing capabilities or collaborate with reputable third-party laboratories.

The implementation of Good Manufacturing Practice (GMP) principles further fortifies the manufacturing process. These principles govern various aspects, including cleanliness, sanitation, equipment maintenance, quality control, and record keeping. By adhering to GMP standards, manufacturers can minimize contamination risks, ensure consistent product quality, and establish a foundation for regulatory compliance.

Consistency and accuracy in manufacturing equipment are vital elements in achieving reliable results. Regular calibration and maintenance protocols are established for equipment such as extraction machines, analytical instruments, and formulation tools. Through such measures, manufacturers can reduce variability and preserve product consistency over time.

A well-trained and educated workforce is essential in upholding quality standards. Comprehensive training programs equip manufacturing personnel with the knowledge and understanding of following protocols, maintaining cleanliness, and adhering to quality control measures. The training covers a range of topics, including safety procedures, equipment operation, formulation techniques, and compliance with regulations.

Detailed and accurate documentation is crucial in ensuring product consistency and complying with regulations. Cultivation and manufacturing processes, ingredient and batch tracking, testing results, and any deviations or corrective actions taken are meticulously recorded. Well-maintained records facilitate traceability, quality control, and regulatory compliance.

To remain competitive and continuously improve, manufacturers must stay updated with the latest research, technologies, and industry's best practices. Active participation in research and development efforts, seeking feedback from customers and patients, and monitoring emerging trends and regulatory changes are essential steps. Embracing innovative techniques enables manufacturers to enhance product consistency and overall quality, solidifying their position in the *Cannabis* market.

Research

The medical *Cannabis* industry can benefit from further research in various areas to improve patient care. Here are some key areas of research that would contribute to advancing knowledge and enhancing patient care in the medical *Cannabis* field.

Conducting robust clinical trials and observational studies to evaluate the efficacy and safety of medical *Cannabis* for different medical conditions is crucial. Research should focus on determining optimal dosage regimens, identifying potential drug interactions, assessing short-term and long-term side effects, and evaluating the risks and benefits of different *Cannabis* strains and delivery methods.

Developing evidence-based treatment guidelines and protocols is essential for healthcare providers to make informed decisions when recommending medical *Cannabis* to patients. Research should aim to establish standardized protocols for specific medical conditions, including dosage recommendations, strain selection, and monitoring parameters. These guidelines will help healthcare professionals provide consistent and effective care.

Further research is needed to understand how cannabinoids and other compounds in *Cannabis* are metabolized and interact with the human body. Studying the pharmacokinetics (absorption, distribution, metabolism, and elimination) and pharmacodynamics (mechanisms of action) of different cannabinoids can provide insights into their therapeutic effects and guide dosage optimization.

Investigating the underlying mechanisms of action of cannabinoids and their interactions with the endocannabinoid system and other physiological systems will help uncover the therapeutic potential of medical *Cannabis*. Research should explore how cannabinoids modulate neurotransmitter systems, inflammation, pain perception, immune response, and other biological processes relevant to various medical conditions.

Focusing on research for specific medical conditions can provide targeted insights into the efficacy of medical *Cannabis*. Conditions such as chronic pain, epilepsy, multiple sclerosis, cancer-related symptoms, inflammatory bowel disease, and psychiatric disorders could benefit from further clinical studies to establish the effectiveness, safety, and optimal usage of *Cannabis*-based treatments.

Personalized medicine approaches within the medical *Cannabis* field will help identify biomarkers, genetic factors, and other patient-specific characteristics that influence treatment outcomes. Understanding how individual variations in genetics, metabolism, and other factors impact response to medical *Cannabis* can facilitate tailored treatment plans for patients.

Research should explore potential drug interactions between medical *Cannabis* and other medications commonly used by patients. Understanding how *Cannabis* compounds interact with pharmaceutical drugs will help identify potential risks and guide healthcare providers in making informed decisions about treatment combinations.

Conducting studies that capture patient-reported outcomes, including quality of life, symptom relief, treatment satisfaction, and adverse events, is vital. These studies can provide valuable insights into the real-world experiences of patients using medical *Cannabis* and help shape treatment approaches and patient support programs.

With the vast size and range of business in the *Cannabis* industry, and services and vendors that serve the industry, many innovative opportunities present themselves when regulatory issues are addressed and resolved.

Cannabis: Not a single industry

Currently, the vastness of the industry attracts many consultants, services, as well as equipment and machine companies. Beverage, tobacco, pharmaceutical companies, and producers of edibles are watching the industry, and some have already started to dabble within it. A few examples are Jazz Pharmaceuticals, Inc. with Epidiolex, along with other pharmaceutical companies. Many beverage and alcohol companies have also entered the market.

Other industries such as beverage, edibles, cosmetics, and skin care are also monitoring the *Cannabis* industry to determine when to join or leave the industry or with whom they will partner. Figure 7 illustrates how mainstream brands have expressed interest in the *Cannabis* market, incorporating it in the infused-beverage market [8] and as discussed earlier, these businesses may be an opportunity for the *Cannabis* industry to collaborate and expand.

As illustrated, the existing *Cannabis* market is clearly not standalone. The industry, as it currently exists, is made up of *Cannabis* companies as well as many ancillary business services and vendors. Within the industry, there are multi-state operators (MSOs), regional *Cannabis* businesses, and independent providers.

As regulations and the industry advance, we should expect to see an expansion of global, multistate, as well as independent business growth. In addition to the current ancillary businesses, there will also be outlier industries that will look to diversify their existing offerings by adding *Cannabis* to their manufacturing and dispensing operations. Several of the industries to most likely diversify are those noted earlier in Figure 7.

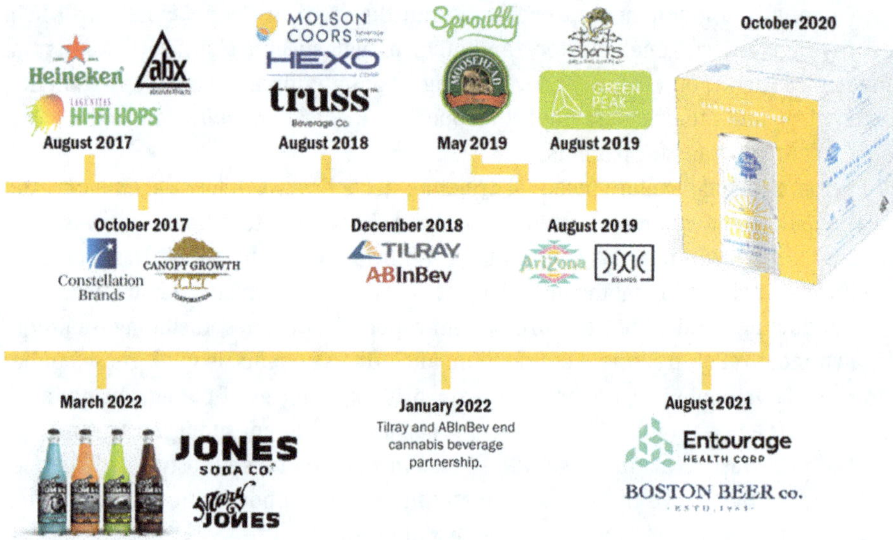

Figure 7: Mainstream Brands in the Infused-Beverage Market.
Source: MJBizDaily research.

Medical *Cannabis* may become a specialized business, similar to other areas of specialization such as pharmaceutical, pharmacy, and research. Recreational marijuana (also known as adult use), as well as CBD products, will see several of the other industries in Figure 7 join manufacturing, processing, and transportation of products such as beer, wine, liquor, coffee, snacks, candy, and tobacco.

Therefore, it will be imperative to lead the industry in innovation, quality assurance and improvement, and best practices and further differentiate the business to better meet the needs of patients and adult-use consumers.

In summary, legislative and regulatory outcomes will determine to what degree the *Cannabis* industry remains standalone and which areas of the industry become components of other industries. Additionally, many other industries may create opportunities to increase their focus on accomplishing solutions to the broader opportunities of the *Cannabis* market.

Clearly, the future holds exciting opportunities as well as significant challenges with competition. Thus, owners must remain passionate, determined, persistent, and constantly focused on making a difference, while improving outcomes and cash flow, in order to be successful in this ever-changing and expanding *Cannabis* industry.

Bibliography

[1] MJBizDaily – Factbook. Legalized States. Lakewood, Colorado, USA. 2022.

[2] State Medical Cannabis Laws. National Conference of State Legislators, 2023. (Accessed August 15, 2023, at www.ncsl.org/health/state-medical-cannabis-laws)

[3] MJBizDaily – Factbook. U.S. Cannabis Sales Vs. Other Industries. Lakewood, Colorado, USA. 2022.

[4] MJBizDaily – Factbook. U.S. Cannabis Industry Employment Comparisons 2021. Lakewood, Colorado, USA. 2022.

[5] MJBizDaily – Factbook. Estimated Number of U.S. Cannabis Businesses in 2022. Lakewood, Colorado, USA. 2022.

[6] MJBizDaily – Factbook. U.S. Cannabis Retail Sales Estimates: 2021-26. Lakewood, Colorado, USA. 2022.

[7] MJBizDaily – Factbook. Timeline of Federal Reform Legislation. Lakewood, Colorado, USA. 2022.

[8] MJBizDaily – Factbook. Mainstream Brands in the Infused-Beverage Market. Lakewood, Colorado, USA. 2022.

Sarah Trent, Seth Tipton

The constant change of business risks

Across all industries, businesses succeed when they adapt to changes and keep up with the pace of change – whether that be statutory, technological, or cultural. This is especially true of the burgeoning *Cannabis* industry. Humans have used *Cannabis* as a medicine for thousands of years across many ethnicities and cultures. In the United States, sentiment about its use changed over the course of the 1900s, culminating during the Nixon presidency with the designation of *Cannabis* as a Schedule I drug, making the plant impossible to research and illegal to possess. Then, with the discovery of the potent beneficial effects of the endocannabinoid system by Rafael Mechoulam in 1988 and the worsening of the opioid epidemic in the United States, the environment became ripe for change that we are seeing the effects of today. Business owners must navigate an ever-changing web of different statutes, codes, ordinances, resolutions, and perceptions in order to maintain compliance and stay ahead of the competition. This chapter examines various risks that *Cannabis* businesses face during the lifecycle of the business – from obtaining the first license to postharvest testing.

Prelicensing approval

Multiple skill sets are required to successfully navigate obtaining the license from the state to operate a *Cannabis* business – ranging from making accurate financial projections to public speaking at town hall meetings. Figure 1 demonstrates the life cycle of a pre-operational *Cannabis* business. Public sentiment about *Cannabis* is slowly changing, but more often than not business owners must be prepared to explain the nuances that go along with operating in the space and be capable of speaking to how *Cannabis* can be properly regulated. *Cannabis* businesses must be resilient, whether in terms of producing more financing or moving forward after sustaining a loss of real estate. Operating in this new industry is not for the faint of heart.

Applying for the license, prelicense capital

In many jurisdictions that permit either adult-use or medical *Cannabis* in the United States, companies that intend to enter the market need to acquire a specific license from a state regulatory agency before they can conduct any business or earn any rev-

Sarah Trent, Valley Wellness, a New Jersey
Seth Tipton, Partner at Florio Perrucci Steinhardt Cappelli & Tipton LLC

https://doi.org/10.1515/9783110677508-005

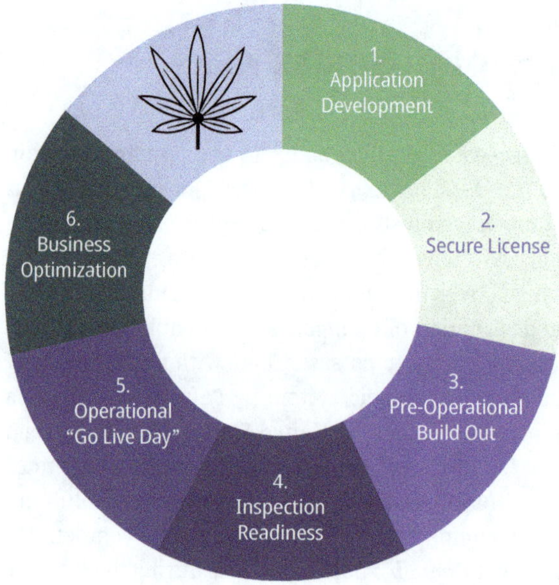

Figure 1: The Business Cycle of the *Cannabis* Industry.

enue. The licensing process can range from a permissive, pass-fail system like New Jersey's present adult use permitting, to a highly competitive, scored merit-based request for applications, like Pennsylvania's medical *Cannabis* program [1]. In both cases, however, applicants are required to meet certain prelicensing criteria that often include the submission of detailed, narrative standard operating procedures (SOPs), architectural renderings, elevations and floorplans, security plans and proof that an applicant has control of real estate at the time of its application that it can build-out upon award of the license. These requirements create significant financial challenges for businesses, particularly new businesses without revenue.

The fact that the application requirements themselves are expensive is not the only challenge to the licensing process. The application review processes in many states have been largely open-ended and present significant risks to the applicant. The first of these risks is disqualification. Some states have taken an all-or-nothing approach to licensing, disqualifying those applicants that were missing any part of an application, no matter how trivial, leaving that potential applicant out of the pool for review, regardless of the capital invested into the application. In other cases, an application process can be enjoined or challenged by court actions, causing delays, additional costs for applicants with pending applications and the possibility that license awards may be overturned. The New Jersey 2019 request for applications was enjoined by court order for one and one-half years as a result of a series of disqualifications made by the agency reviewing applications [2]. The hurdles that face *Cannabis* businesses from the start almost all require significant capital to address, and *Cannabis* business owners must be raising money at all times.

Company valuation, raising money

Cannabis companies cannot resort to traditional banking and commercial loan options when raising the funds necessary to participate in the application process, licensing and commercial opening. As a result, one of the first and perhaps most challenging aspects of starting and operating a *Cannabis* business is that the owners must be constantly fundraising from private equity sources to meet the intense capital demands of licensing and operations. Before raising any significant amount of money, however, a *Cannabis* company must identify its value to investors. This is an imperfect and imprecise process for a number of reasons.

A primary consideration in valuing a *Cannabis* company is evaluating the licensing status of the state in which the company operates. In an open-licensing state such as Colorado or Massachusetts where an applicant may qualify for a license simply by submitting a compliant application, a license may be worth far less than in more competitive states. Conversely in state that have statutorily capped the number of licenses that are allowed to be issued, such as Pennsylvania, licenses can be valued in tens of millions of dollars. Licenses can also have significant features that can increase or decrease the value of the license, such as limitations on the amount of cultivation canopy sizes, or limits on the amount of product a retail licensee can sell (i.e., New Jersey microbusiness licenses). Another element of valuation requires estimates of customers and revenues. This can be a relatively objective process in states with developed programs and historical sales, like Colorado or California. However, in states with brand new licenses and no historical sales, coming up with a realistic financial model is difficult if not impossible to do with any real certainty.

In addition to license-value considerations, location can be critical. In a state that has a liberal licensing program, the value of the licensee's local approvals may be determinative in valuing the company. For example, in several of New Jersey's urban areas, such as Hoboken and Jersey City, the cities engaged in competitive processes to select the licensees that would be permitted to operate there. While the state license may have little value as a result of how many have been issued, the local approval could significantly increase a company's financial worth.

Real estate and landlords

A *Cannabis* company that has raised enough capital to initiate the application process is typically required to identify to the licensing agency where that applicant will operate when and if a license is granted. The licensing agencies typically require some form of evidence that the applicant can actually operate at a given location, usually in the form of a lease, letter of intent, option agreement, or deed. The so-called "proof of control" is a concept that is a constant challenge for almost all *Cannabis* businesses.

The primary issue that this presents is cost. A *Cannabis* company intending to open a dispensary typically wants to open at an ideal retail location. However, at the time of securing the location, the same *Cannabis* company probably does not have revenue, and cannot tell the landlord with certainty when a license might actually be granted by a state agency. A landlord with a desirable retail location is not likely to agree to leave its space vacant and bet on a *Cannabis* company becoming licensed when other traditional retail stores may take occupancy and begin paying rent far faster. As a result, particularly in more urban settings with competitive real estate markets, *Cannabis* companies must budget for paying a full rental rate for all, or at least a portion of the period during which their application is pending, and their subsequent build-out.

Once a *Cannabis* business has identified a location and perhaps reached an agreement to secure that real estate during the application process, there are other risks that can arise to undermine their plans. For *Cannabis* tenants, one problem that frequently arises is the landlord's lender. Most traditional lenders are unwilling to do business with *Cannabis* businesses and do not make loans to commercial real estate owners that are leasing to *Cannabis* businesses [3]. As a result, whether through routine diligence, lease approval, or a request for a subordination and nondisturbance agreement (SNDA), lenders learn of *Cannabis* business tenants and then frequently either bar the lease entirely or force the landlord to terminate the lease or refinance. Moreover, because the lenders will not do business with *Cannabis* businesses, they are typically not offered any tenant improvement allowance to offset the cost of improvements and have little protection from a foreclosure on the landlord because lenders will not enter SNDAs with *Cannabis* business tenants [4].

Town approvals, permits, postlicense spend

Local public opinion on *Cannabis* businesses and the politics surrounding establishing a *Cannabis* business are perennial issues that can change rapidly. State licensing agencies have recognized that a critical component to awarding a license is confirming that the applicant has the local approval to open at its stated location. As a result, in many jurisdictions, such as Colorado, Massachusetts, and New Jersey, licensing agencies require proof in various forms that the applicant has approval to operate a *Cannabis* business at the location it controls, whether from the zoning board, a planning agency, or the governing body itself, such as a town council or committee. How a *Cannabis* business goes about obtaining that approval can be time-consuming, extraordinarily competitive, and very expensive.

The local approval process generally has two main subsets: those that occur prior to a license application, and those that are conducted after a license is awarded. Both present serious concerns to *Cannabis* businesses. A licensing process that occurs prior to a state-level application will typically require the applicant to commit some capital to identifying and securing real estate, and then drafting and submitting a compliant local application. If that applicant is unsuccessful in the local process, the *Cannabis* business may be effectively blocked from submitting a state application by the lack of a local approval, and with a substantial loss of capital.

A licensing process that occurs after state licenses are granted is even more potentially fraught for *Cannabis* businesses. In some jurisdictions, the local municipality or city will permit all applicants to receive basic, preliminary approval, allowing the applicant to make an application to the state agency, but reserving limited, local licenses for those businesses that successfully obtain a state license. Applicants seeking to establish in a location like this must take all of the cost, time, and risk of identifying and controlling real estate, and submitting a state application, while knowing that their local approval is not guaranteed. This becomes even more complicated when a city or municipality takes an approach that the first applicant (versus the most experienced applicant) should be awarded a local license, based merely upon such applicant being the first to submit an application. Given the many contingencies involved in the state licensing processes, the first to obtain a state license can come down to luck, forcing the business to put serious investor capital at risk with little or no certainty as to an outcome.

Postlicensing approval

The risks and challenges of establishing a *Cannabis* business are not over when a company has obtained a license, built out its facility, and opened its doors for business – they are just starting. For those that have never operated a *Cannabis* business before, the unknowns are endless: will the point of sale (POS) function properly and integrate with other necessary software, how to get employees badged, how will 280E really effect my business, and what should go into filing your first tax returns are all questions that will need to be answered along the way. Retailer, cultivators, and manufacturers will all face unique hurdles on the path toward business optimization.

Design and construction

Layout of the facility and building materials are some of the first things decided in the postlicense phase. In order to get construction underway, the business needs a full set of construction drawings, which can take months to produce. Choosing between different types of fixtures, equipment, and materials will likely be daunting to a first-time business operator compared to a multistate operator, designing a retail shop for the umpteenth time, with fixtures and materials likely already chosen from prior projects. Businesses must have a good working relationship with all professionals on the project, including, architects and engineers. Many times, a project will only move forward as quickly as they can work; most construction materials will require shop drawings, which will need to be approved by either the architect or engineer on the project, in order to ensure that the facility is being built to code. To avoid costly change orders down the road, businesses should take time from the outset of the project to ensure that the workflow concept matches with the architectural design, including consideration of what POS and how orders might be filled. Vetting POS companies and comparing quotes take time, forcing business owners to be working on multiple tasks to facilitate proper facility design.

It is also important to allow the regulatory body time to review the construction documents and provide comments in order to avoid costly mistakes later. For instance, many states require that the product storage vault be made out of certain materials or built to certain specifications, all the way up to requiring a secure vault built to comply with the Code of Federal Regulations (CFR) [5]. There could be different storage requirement for medical *Cannabis* versus adult use *Cannabis*, and any deviations from the statute or code might require a waiver from the regulatory body.

Security – People, product, cash

Startup *Cannabis* businesses should not underestimate the cost and infrastructure that a new security system will entail. Equipment and guards are necessary to protect the businesses assets and staff but are also required by the regulatory body. A properly working system will be required before final licensing. Security infrastructure will inform early decisions of the business, some of which may also need to be approved by the regulatory body, such as layout of the facility, the vault or cage, and what software is used to allow the remote access of the surveillance equipment. Security protocols or a security manual can be found in most *Cannabis* business. These procedures are required and approved by the regulatory body and are part of the businesses obligation to maintain in order to stay compliant.

Video surveillance is the centerpiece of the *Cannabis* business security system. Every part of the facility where *Cannabis* or *Cannabis* products are grown, prepared, or stored must be under video surveillance in order to stay compliant. In addition to product security, staffs' safety is paramount. Other considerations for staff and customer safety include parking and exterior building lighting to ensure a safe environment to get to and from the business.

Being able to view the security feed in real time, remotely, from a smartphone or portable device assists business owners working remotely and allows the regulatory body access to the security feed. In some jurisdictions, the regulatory body has real-time access to the video surveillance system. For instance, a regulator may want to ensure that a specific standard operating procedure was followed when the company disposed of *Cannabis* waste, so they log-in to watch the video of the destruction that happened in the past.

Procedures to avoid product and cash diversion are typically submitted to the regulatory body during the licensing review process. While cultivators must find ways to incentivize staff not to divert product before packaging, on the retail side, it is key to make sure that cash is accounted for and reconciled regularly, and that staff are properly trained on cash handling and counting procedures. Video surveillance allows for accountability on all fronts.

Security guards can be found in many *Cannabis* businesses, specifically retailers. Security guards could be required by state or local law. In some cases, the locality will require a guard to limit the possibility that the local police are called. Guards, whether armed or unarmed, provide security to staff and serve as a deterrent to anyone targeting the business. Cultivators and manufacturers have a lot of products sitting on hand, and employees coming to and from work – so even on large cultivation work sites, there will likely be a security guard.

Connecting across the supply continuum

Finding the right equipment and software that is right for the business is not new to the *Cannabis* industry, but for a first-time *Cannabis* business owner choosing the "tech stack" can become overwhelming. One of the biggest decisions is choosing the best POS or seed to sale tracking system to use. Most states with a regulated *Cannabis* market mandate the use of METRC, a seed to sale tracking system that allows the State to view inventory and shipments in real time. However, METRC does not have all the functionality that other POS software does, so operators use METRC (if mandated by the Regulatory body) in combination with other POS or seed to sale tracking systems, like Dutchie, BioTrack THC, or MJ Freeway, to name a few. Ensuring that the POS has built in capabilities to assist in implementation of the specific business model is key. Retailers, for instance, want to determine how orders will be filled (will the business operate a kiosk/fulfillment model or deli-style model where product is kept out on the sales floor). Reporting and the ability to generate a variety of very specific reports, along with the ability to integrate with accounting software such as QuickBooks, are some of the features that can set a good POS apart from others.

Each aspect of the supply chain has specific considerations to take into account. For instance, cultivators may look for ease of functionality in tagging and moving plants between rooms. Retailers, on the other hand, are going to be interested in the ease of which other software integrates with the POS. For optimal operations, different facets of the business should integrate with the POS. Integration refers to the ability for two pieces of software to share information seamlessly through an API key or Application Programming Interface. When software does not integrate properly, the business will incur extra costs by having to manually transfer/upload/download information between different programs. The customer relations management ("CRM") platform, menu, ordering, banking, Weedmaps and Leafly, and any payment platform should integrate for optimization of the business.

A retailer's tech stack may include home delivery software, programs that allow the menu to be displayed on TV screens, software to track and run employee incentive programs, employee scheduling tools, sales reporting tools, security interfaces that provide real-time alerts, just to name a few. Then, on the cultivation side, there are a number of options to automate irrigation and fertigation, allowing growers to remotely control what is happening inside of the facility. Finally, commercial *Cannabis* manufacturing operations depend on sophisticated pieces of equipment and choosing the right piece of equipment for the scale and goal of the operation is key. The industry is in its infancy; therefore, we will continue to see innovations in technology that make all aspects of plant touching operations more convenient, safer, with diversification of products.

A variety of payment solutions integrate with the retailers POS, but those options also need to be compliant with bank and the way that it accepts deposits. Increasingly banks are doing business with plant touching *Cannabis* companies as they find ways to help limit their liability by establishing extra checks to ensure that clients are not laundering money. For instance, some banks mandate the use of Green Check Verified, yet another piece of software that integrates with the POS to ensure that deposits match what is happening at the POS.

Banking/Section 280E

Banks providing banking services to *Cannabis* businesses are facilitating violations of the Controlled Substances Act. As a result, according to the federal government as recently as cases litigated by the Justice Department in 2017 [6] and a memo from the Justice Department known as the 2014 Cole Memo [7], those banking services are a direct violation of the Bank Secrecy Act[8], which prohibits banks from allowing criminals from accessing and utilizing the federal banking system. Moreover, the federal Money Laundering Control Act[9] defines the proceeds of *Cannabis* sales prohibited under the Controlled Substances Act as the proceeds of illegal money laundering that is off limits to most banks. For these reasons, traditional commercial loans are almost nonexistent for traditional banks, and deposit relationships often must be formed with state-chartered banks and require the payment of exorbitant fees.

In addition to the problems that arise from maintaining a traditional banking relationship for a *Cannabis* business, many are shocked after their first year of operations to learn of the impacts of Section 280E (26 U.S.C. 280E) in the Internal Revenue Service (IRS) code. Companies are typically permitted under the IRS code to deduct from their income the cost of carrying on their business and, as a result, reduce their taxable income. In 1982, however, a United States Tax Court issued an opinion in *Ed-*

monson v. Commissioner in which the court permitted a drug dealer to deduct the cost of traveling across the country to purchase amphetamines, cocaine and marijuana, as well as his cost to purchase the illegal drugs. This alarmed Congress, which immediately drafted Section 280E, which flatly prohibited any deductions for "carrying on any trade or business, if such trade or business . . . consists of trafficking in controlled substances (within the meaning of Schedule I and II of the Controlled Substances Act)." Regardless of state licensure, *Cannabis* businesses necessarily traffic in a Schedule I substance. Consequently, with very few exceptions, courts have routinely held that *Cannabis* businesses may not deduct from their income numerous different expenses that any other company would be permitted to deduct, leading to a higher tax burden for those companies.

Compliance and state regulators

Compliance in this highly regulated industry is key. Everything is on video and recorded, and written logs and electronic databases store all company documents dating back years. In some regulated markets, the SOPs of the business are approved by the regulatory body before issuance of the permit, allowing for imposition of fines and penalties in instances of noncompliance.

A good working relationship with the regulatory body and any compliance monitor that may be assigned to the business is key. In many cases, communication and documentation of all goings-on inside of the facility is the starting point for that working relationship with the compliance monitor. For instance, communicating a mistake that was made – while mistakes should be avoided, reporting those mistakes in a timely manner can gain the company credibility. Companies employ people and people make mistakes – what matters is how those mistakes are handled and reported.

There are a few things that are at the top of the compliance list, including advertising, packaging and labeling, and testing. Meaning that a violation of a provision relating to one of these topics could mean imposition of a fine, or worse. Advertising rules apply to all *Cannabis* companies – cultivators, manufacturers, and retailers, and while rules vary from state to state, advertising is also regulated at a local level. For instance, cities and municipalities control and approve signage and exterior building façade plans, and usually want to ensure that *Cannabis* leaves are not displayed on signs and that paraphernalia is not visible through the windows to passersby. Submitting all advertising and signage plans to the regulatory body and to the local permitting authority is important.

Compliance in packaging and labeling is just as important as knowing the advertising rules inside and out, although fines are less likely because packaging and labels are more likely to be specifically approved for by the regulatory body before the customer sees it. In other words, businesses might have to submit all packaging designs for approval before putting that product/packaging out for sale, a process that can take time depending on the resources behind that particular regulatory body. Most states mandate the use of a warning symbol (like below, from New Jersey) that is applied to all *Cannabis* packaging, in a very specific size and color. Figure 2 is the mandatory warning label that must appear on all *Cannabis* packaging in New Jersey; failure to include this mark in the correct size on any *Cannabis* packaging could result in a warning or fine from the regulatory body. It is possible that as the market matures, and industry associations and state task forces begin to share information and work together, that a universal warning label could be adopted across the industry. On the retail side, patient labels may also have to be approved for use by the regulatory body, and there are likely rules that require a patient label to be applied to every product dispensed to a qualifying patient. That label needs to contain very specific information – like the business license number and address that dispensed the product (which might be different than the business that manufactured the product).

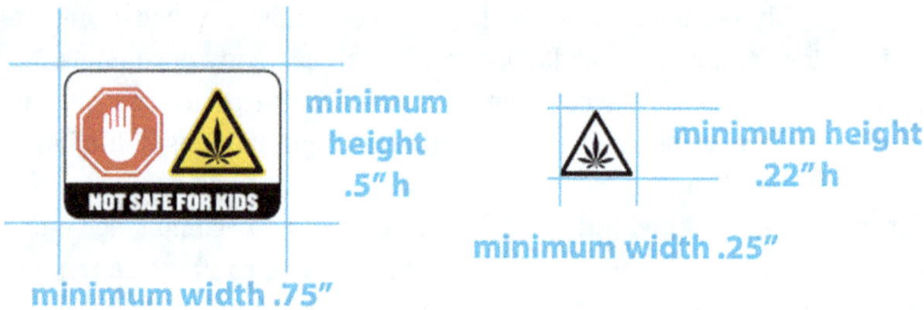

Figure 2: *Cannabis* Warning Labels Required on all *Cannabis* Packaging in New Jersey.

Many states already require detailed warnings and information about source, ingredients, and testing to be directly on the product label. Cultivators and manufacturers must ensure that products fall within proper testing limits, otherwise can face destruction of those products. Patients and consumers want to know what a product's THC or CBD level is, but testing goes far beyond that. It ensures that cultivators are not using prohibited pesticides and that flower is free from mold and heavy metals. On the manufacturing side, it ensures that prohibited ingredients never make their way into a vape cart, for instance.

A final point to make in talking about compliance is the difference between the medical market and serving patients that have qualifying conditions versus the adult use or recreational market as it relates to the businesses ability to prioritize patients, especially if it is required under the law. Some markets may require a separate entrance for medical patients, separate registers, and even separate products, all in an attempt to ensure that medical *Cannabis* patients do not wait in line or run out of a diversity of products.

Crop failure, pesticides, testing

Under the right conditions, *Cannabis* is a crop that grows very easily – hence, the name "ditch weed." But in reality, commercial *Cannabis* cultivation is complicated, with many variables to be taken into account. Although outdoor grow in California may be able to take down more than one harvest a year, an outdoor cultivator on the East Coast faces intense mold and mildew during hot and humid summers and cannot get more than one harvest per year due to temperature and lighting. Pests, such as the corn earworm and corn borer (Figure 3), cause devasting damage to an outdoor crop rendering it nearly useless, and some states, like New Jersey, have very strict prohibitions on the use of pesticides on *Cannabis*. Many of the risks related to pests are decreased by growing indoors with the ability to control the environment.

Figure 3: Picture of Bud Rot Caused by a Corn Borer, Taken by Sarah Trent, Long Valley, NJ, October 2021.

Understanding the intended end use of the product will inform how that product is treated throughout its life cycle. *Cannabis* flower from top genetics, grown indoors with little pest pressure (pests range from insects to fungi, see Figure 4), can be sold in medical and adult use dispensaries for a premium. If *Cannabis* flower does not test as well as expected, a cultivator might decide to send that batch for extraction instead of making the flower available for retail sale. In part, this is because some extraction processes "clean" a product where pest pressure was high. Furthermore, products generated from extraction can be mixed with other products to make up for deficiencies, such as the common practice of adding artificial terpenes to manufactured products, instead of using the terpenes that were pulled from the *Cannabis* plant.

Figure 4: Picture of Moth Eggs, Taken by Sarah Trent, Long Valley, NJ, October 2021.

With testing now mandatory in all medical and adult use markets, patients and consumers have the ability to make informed choices at the dispensary instead of relying on generic terms like indica and sativa. In many markets, testing and labeling requirements now allow patients to examine the precise cannabinoid and terpene profiles of a product. However, testing requirements are a point of contention between different groups within the industry. Although cultivators attempt to sway regulators toward larger batch sizes, therefore less testing, other stake holders may like to see more testing, depending on the market and its maturity.

Conclusion

For many in America, *Cannabis* business can be seen as a business in which millionaires are minted overnight as states move toward legalization. However, *Cannabis* companies have all of the same challenges as that of traditional start-ups, but with a host of additional challenges as outlined in this chapter relating to its current status as a federally illegal product. Any person attempting to enter into the industry should carefully consider each of the factors and challenges, whether it be for funding, the choice of where to operate, or the inherent risks in sophisticated horticultural operations.

References

[1] New Jersey *Cannabis* Regulatory Commission, "Notice of Application Acceptance for Personal-Use *Cannabis* Business Licenses," dated July 18, 2023, available at https://www.nj.gov/*Cannabis*/documents/businesses/personal-use/Final%20Notice%20of%20Application%20Acceptance.pdf (last visited September 13, 2023) (describing the pass-fail scoring procedure for review of applications); *see also* Keystone ReLeaf LLC v. Pennsylvania Dept. of Health, 186 A.3d 505, 509 (Cmw. Ct. of Pa. 2018) (describing competitive scoring process and number of applications received for limited number of licenses).

[2] In re Medicinal Marijuana Alternative Treatment Center for Tetra Grow LLC, 2021 WL 631238 (Docket No. 1-1275-19) (N.J. Super. Ct. Feb. 18, 2021) (vacating fifteen month stay).

[3] Taylor DB, "Mastercard Moves to Stop Use of Debit Cards at *Cannabis* Shops," The New York Times (July 28, 2023) (noting that "most big banks and credit providers will not work with marijuana businesses" because "federal law prohibits the sale, possession, and use of marijuana in all of its forms.").

[4] K&L Gates LLP. "Tenant's Use – Illegal or Not?", JD Supra (March 31, 2014) (Available at https://www.jdsupra.com/legalnews/tenants-use-illegal-or-not-59066/) (last visited Sept. 13, 2023) ("A request for a subordination and non-disturbance agreement (SNDA) is likely to become a significant obstacle for the lease transaction and may potentially kill the deal, as a federally insured and regulated lender may object to a use of the property that is illegal under federal law.").

[5] CFR. Title 21, Chapter 2, § 1301.71 Security requirements generally.

[6] Fourth Corner Credit Union v. Fed. Reserve Bank of Kansas City, 861 F.3d 1052 (10th Cir 2017).

[7] United States Justice Department, Office of the Attorney General, "Guidance Regarding Marijuana Related Financial Crimes," (Feb. 14, 2014).

[8] 31 U.S.C. 5311 et seq.

[9] 18 U.S.C. 1956 and 1957.

Kevin Fortin, Ruth Charbonneau

Intellectual property in the *Cannabis* industry

The *Cannabis* industry in the United States is fragmented due to diverging principles of federal and state laws governing marijuana. Under the Controlled Substances Act (CSA)[1] enacted in 1970, "marihuana" and "marijuana" mean all parts of the plant *Cannabis* sativa L., growing or not; the seeds thereof; the resin extracted from any part of such plant; and every compound, manufacture, salt, derivative, mixture, or preparation of such plant, its seeds or resin.[2] Marijuana is listed as a Schedule I controlled substance under the federal CSA, meaning that it has a high potential for abuse, no currently accepted medical use in treatment in the United States, and a lack of accepted safety for use under medical supervision. In 1996, in part as a response to the AIDS epidemic, California voters passed the Compassionate Use Act, allowing for the use of medical *Cannabis* and at least at a state level rejected the federal Schedule I designation. During this same period, general population surveys reflected increasing support for legalization of medical *Cannabis* as state legislatures debated changes to reduce penalties for *Cannabis* offenses and to legalize the use of medical *Cannabis*. The U.S. Department of Justice in 2009 issued the first of several memos advising U.S. Attorneys "to not focus federal resources in [their] States on individuals whose actions are in clear and unambiguous compliance with existing state laws providing for the medical use of marijuana." In 2012, Washington and Colorado approved the use of *Cannabis* for medical and adult (nontherapeutic) use. In 2018, the CSA was amended by the Farm Bill[3] to remove hemp from the CSA by defining it as the plant "*Cannabis* sativa L. and any part of that plant, including the seeds thereof and all derivatives, extracts, cannabinoids, isomers, acids, salts, and salts of isomers, whether growing or not, with a delta-9 tetrahydrocannabinol concentration of not more than 0.3% on a dry weight basis." Federal laws and regulations have had little impact on efforts by individual states to legalize and decriminalize the use of marijuana. The primary rules governing these programs arise from each state's legal and regulatory scheme. As of the writing of this text, 37 states have legalized the medical use of *Cannabis* to varying degrees and 23 states have medical and adult, or nontherapeutic, use.[4]

The meaning of the terms "*Cannabis*" and "Cannabinoid" vary between states, further muddying the federal-to-state and state-to-state waters. In California, the statutory definition of "*Cannabis*" expressly includes what is commonly understood to be

1 21 U.S.C. 801 et seq.

2 21 USC § 802(16).

3 Agriculture Improvement Act of 2018, P.L. 115–334.

4 This does not include the District of Columbia or U.S. territories.

Kevin Fortin, Fortin Patent Firm
Ruth Charbonneau, Thomas Jefferson University

https://doi.org/10.1515/9783110677508-006

marijuana, but specifically does not include industrial hemp. Cannabidiol (CBD) is a chemical constituent of the *Cannabis* plant that is encompassed within the CSA's definition of marijuana. Unless stated otherwise, the usage of the term "*Cannabis*" herein is intended to be the botanical definition.

Introduction to intellectual property landscape

The burgeoning *Cannabis* industry is rife with innovation, and as it continues to grow, so do the opportunities and challenges surrounding intellectual property (IP) rights. This chapter explores the diverse landscape of IP within the *Cannabis* industry, delving into patents, trademarks, copyrights, and trade secrets. We'll also examine the intricate web of federal laws and regulations that intersect with the *Cannabis* industry and discuss innovative practices and opportunities within this unique space. The 2018 Farm Bill explicitly preserved FDA's authority to regulate products containing *Cannabis* or *Cannabis*-derived compounds under the Federal Food, Drug, and Cosmetic Act (FDCA). Registration of marks for foods, beverages, dietary supplements, or pet treats containing CBD will still be refused as unlawful under the FDCA, even if derived from hemp, as such goods may not be introduced lawfully into interstate commerce.

At the same time, the federal government has readily issued utility patents on legal *Cannabis* inventions. The primary right accompanying an issued patent is the right to exclude others from practicing it – which generally requires the patent owner to sue in federal court. As a result, the enforceability of legal *Cannabis* patents directly impacts their value.

Patents

Patent law became popular in the United States during the industrial revolution. At the time, land patents were also popular, being granted to settlers moving from crowded east coast conditions to the western United States. In addition to land patents, miners and hobbyists could "claim" mining rights by staking an active mining claim and working it productively. The staked claim was then registered at the local land patent office.

A U.S. patent gives you, the inventor, the right to "exclude others from making, using, offering for sale, or selling" an invention or "importing" it into the United States. A plant patent gives you additional rights on the "parts" of plants (e.g., a plant patent on an apple variety would include rights on the apples from the plant variety). What is granted is not the right to make, use, offer for sale, sell, or import the invention, but the right to stop others from doing so. If someone infringes on your patent, you may initiate legal action. U.S. patents are effective only within the United States and its territories and possessions.

Three types of patents relevant to *Cannabis*: Utility, design, and plant

Utility patents are the most common type of patent in the *Cannabis* industry. They protect new and useful *Cannabis*-related inventions, such as novel cultivation methods, extraction processes, and drug delivery systems. *Cannabis*-related innovations can generally obtain utility patents without significant hurdles. At least 90% of patents filed fall into this category.

Design patents are essential for safeguarding the ornamental, nonfunctional, and aesthetic aspects of commercial *Cannabis*-related products, such as unique packaging or the design of *Cannabis*-infused edibles. Due to state administrative rules developed to assure safety of *Cannabis* products or child-safety requirements, numerous creative packaging designs have been developed and promoted. These are typically entitled to design patent protection. A design patent covering a creative-looking packaged product may be a useful alternative to trademark protection. Design patents are limited in duration to 15 years. Design patent protection for packaged products is underutilized and may be a cost-effective alternative to trademark protection in some cases.

A plant patent provides a unique form of IP protection for new and distinct plant varieties that have been asexually reproduced (i.e., propagation by cloning). In the *Cannabis* industry, obtaining a plant patent can be a strategy for protecting proprietary strains and genetic innovations. The United States Patent and Trademark Office (USPTO) will only grant a plant patent if it receives a full and complete botanical description that explains how the plant is unique and includes drawings showing the plant's unique features. An alternative to plant patents is the standard utility patent that can cover the genetic sequence of your *Cannabis* plant, including biomarkers and engineered artifacts. This does not require asexual reproduction and can cover broadly sold *Cannabis* strains. Plant patents cover ornamental features of a plant. They are used in the rare circumstance that a *Cannabis* plant is bred for ornamental purposes, but in the industry this is rare.

However, if a new strain of *Cannabis* was developed to optimize for a particular cannabinoid that is not normally found in abundance, this strain may be patented via a utility patent. If the strain expresses purple and orange stripes on its leaves or buds, then it could also be considered for a plant patent to cover the aesthetic features. For example, if a new type of hemp is engineered so that the superior fiber can be used broadly across the world for textiles, then the developer of the hemp strain may be interested in a utility patent for his/her new hemp strain with superior fiber. This would not be considered for a plant patent, but for a utility patent.

Utility and plant patents have a term for up to 20 years from the date the application was filed, with maintenance fees on a certain schedule to keep the patent in force. Although the patent term is for 20 years, the application becomes publicly available after 18 months. A design patent is granted for a term of 15 years from the date of grant. Under certain unusual conditions, patent terms may be extended or adjusted.

Plant patents are great for yellow and orange striped roses that are sold in large volumes. However, seldom do *Cannabis* plants express super unique looks and colors required to obtain plant patent protection.

Patent licensing arrangements

Patent law conceptually is very similar to real-estate law. Real-estate law has varied attributes, i.e. mineral rights, water rights, easements, etc. Holding a deed to real estate enables one to evict trespassers. Similarly holding an issued patent grants one the right to keep others from making, using, selling, and importing the patented invention. A patent can be licensed, sold, and used for collateral for a loan. Patents can be used for many other purposes including financing further research and development of a company. Thus, patents offer exclusivity in the market place in return for publishing the invention. Patents normally expire after 20 years when the rights thereunder terminate and the invention is ceded to the public domain.

Consider that different people have different skills. Some are good sales people and deal makers, some have an affinity for growing plants, some are good at operating a store front, others are happy in a lab experimenting in solitude. Economic efficiency is gained when people practice their best skills. This same concept applies to companies. Some are innovative, some have manufacturing expertise, and others are good at marketing and expanding markets. Often a small company has a clear strength, but may lack technical innovative expertise, manufacturing savvy, or business development and strategic expertise. This is why licensing technology is so important. Licensing enables innovators and innovative companies to monetize their innovation without having to invest in manufacturing, distribution, and sales.

A stereotypical example of a licensing strategy in the *Cannabis* space includes an innovative company in state X, who licenses its product and manufacturing technology to a large market-savvy operator in state Y. The innovator trains the operator to make and package the innovative product under license in state Y. A moderate royalty is paid, along with consulting fees. The innovator can also sell packaging materials bearing a brand to the operator in state Y.

This licensing arrangement example benefits both parties. The cost of the royalty to the operator in state y is much less than the development expenses. The innovator can also look at other states to engage more licensees. An innovator can easily multiply (e.g., 5×) its licensing revenue by establishing licensing in more states. An operator can easily make and sell innovative products to add value to its existing markets in a single state.

Ideally, the licensor would license many IP rights. The rights could include patent rights to the product and manufacturing method, trade secret rights in product and manufacturing variations, and optimal manufacturing equipment, layout, and control systems. Trademark rights in the brand can be licensed along with quality control methodology. Trade dress rights[5] in the product packaging can be licensed.

While the array of types of rights may be a bit overwhelming at first glance, a skilled IP attorney can easily craft these into a licensing agreement. If a licensing deal falls apart or expires, the agreement can be terminated. Then there are contractual rights, trademark rights, trade secret rights, patent rights, and trade dress rights that can be enforced together or separately as necessary. Even if independent third party is caught using the innovator's technology, the patents can be enforced independently. In exceptional cases of willful infringement, treble damages and an injunction are possible.

It is also possible to enforce patents against any in the supply chain, including against customers of the unlicensed operator and downstream retailers. I have had some interesting discussions with the customers of unlicensed operators about patent infringement. Such discussions can lead to wiser and more strategic sourcing choices that stop the infringement and favor the innovator. Extrajudicial action, if performed skillfully, is really a sophisticated form of salesmanship and business development.

Cannabis industry regulatory nuances

In states where *Cannabis* is legalized for medical or adult use, the industry typically operates under a framework that bypasses many Food and Drug Administration (FDA) regulations, as the FDA primarily governs products intended for medical use and consumption.

In summary, the *Cannabis* industry in states with legalized *Cannabis* operates largely outside of FDA regulations and instead adheres to state-specific regulatory frameworks. See Table 1. The industry's unique legal status presents both opportunities and challenges for businesses and regulators alike.

5 The design and shape of the materials in which a product is packaged. Product configuration, the design and shape of the product itself, may also be considered a form of trade dress.

The Lanham Act protects trade dress if it serves the same source-identifying function as a trademark. It is possible to register trade dress as a trademark, but for practical reasons most trade dress and product configurations are protected without registration under 15 U.S.C. §1125(a). trade dress | Wex | US Law | LII / Legal Information Institute (cornell.edu).

Table 1: Overview of State-Legal *Cannabis* Regulation.

***Cannabis* Industry Operates in States and Outside of the FDA Regulations**	
State-Level Regulations	In states with legal *Cannabis* programs, regulatory agencies, boards, or commissions (agencies) oversee the industry. These agencies create and enforce rules and guidelines specific to the cultivation, processing, testing, packaging, labeling, and distribution of *Cannabis* products. For example, California has the Bureau of *Cannabis* Control, while Colorado has the Marijuana Enforcement Division.
Product Safety and Testing	States with legalized *Cannabis* often require rigorous testing of *Cannabis* products for potency, contaminants, and safety. This testing is typically carried out by independent labs accredited by the state. These state-level testing requirements are often more comprehensive than FDA regulations for other products.
Labeling and Packaging	*Cannabis* products are required to have clear and informative labeling, including information about potency, ingredients, serving sizes, and safety warnings. The labeling and packaging requirements vary from state to state but are typically aimed at ensuring consumer safety and informed choices.
Quality Control	The *Cannabis* industry is responsible for its own quality control and standards, as there is no federal oversight by the FDA. Companies must develop and maintain their quality control processes to meet state requirements and maintain product consistency and safety. These requirements vary state to state.
Advertising and Marketing	*Cannabis* companies are subject to state-level regulations regarding advertising and marketing, including restrictions on advertising to minors and false or misleading claims about products. These regulations are designed to prevent predatory marketing practices.
Patient and Consumer Education	Many states with medical *Cannabis* programs require dispensaries to provide educational resources to patients about the potential risks and benefits of using *Cannabis* for medical purposes. These educational efforts aim to ensure that patients make informed decisions.
Product Variety	*Cannabis* businesses can offer a wide range of products, including flower, edibles, concentrates, topicals, and more. These products cater to both medical and recreational and may not meet FDA standards for pharmaceuticals or dietary supplements. Allowable forms vary by state.
Business Licensing and Compliance	Plant-touching *Cannabis* businesses must obtain licenses from state regulatory agencies, and compliance with state regulations is mandatory. Failure to comply can result in penalties, fines, or the loss of a license.
Taxation and Revenue	States generate significant tax revenue from legalized *Cannabis* sales. Tax rates and structures vary widely, and these revenues are typically used to fund various state programs and initiatives to support social equity applicants and communities.

Trademarks and trade secrets

After the passage of the 2018 Farm Bill legalizing hemp and its derivatives, the USPTO lifted its ban on registering *Cannabis*-related marks, creating an exception allowing registration of marks related to hemp and certain types of CBD. USPTO noted:

> [f]or applications filed on or after December 20, 2018 that identify goods encompassing *Cannabis* or CBD, the 2018 Farm Bill potentially removes the CSA as a ground for refusal of registration, but only if the goods are derived from "hemp." . . . If an applicant's goods are derived from 'hemp' as defined in the 2018 Farm Bill, the identification of goods must specify that they contain less than 0.3% THC. Thus, the scope of the resulting registration will be limited to goods compliant with federal law.[6]

Trade secrets offer an advantage over patent registration in that trade secrets do not expire. But unlike patent protection, protecting an invention as a trade secret does not provide the ability to exclude third parties from use of the trade secret so long as the third party independently discovered or developed the invention through other lawful means.

Regarding trademarks, federal trademark registration is prohibited for marijuana and cannabinoids isolated from hemp under the theory that trademark registration is not available for illegal products, including those lacking FDA approval. A company can invest a substantial sum in marketing nationwide to bolster sales in various states only to face the persistent risk of knock-off manufacturers using the trademarks of others.

Trade secret law, for example, may protect formulations for popular cola beverages. Trade secret law[7] may also protect secrets arising from unique processes applied to extraction and formulation of cannabinoids. Trade secret protection is nuanced because some states require detailed disclosure of processing methods and ingredients in the *Cannabis* industry. These rules are fragmented, varying from state to state. Corporate discipline and consistency are always required to keep a trade secret. The owner must show consistent adherence to protocols designed to protect the trade secrets, such as limiting access to the trade secrets, insisting on non-disclosure agreements, marking documentation, regarding the trade secret as confidential. The term is as long as the trade secret remains protected and does not become publicly known.

The *Cannabis* industry includes many managers, scientists, and technicians who are relatively early in their careers and have never adhered to the strict compliance with corporate policy that is required to enable effective trade secret protection. I have heard of incidents of the disclosure of trade secrets after a late night bong hit on a

6 Exam Guide 1–19.pdf (uspto.gov).

7 In the United States, trade secrets are defined and protected by the Economic Espionage Act of 1996 (outlined in Title 18, Part I, Chapter 90 of the U.S. Code) and also fall under state jurisdiction. As a result of a 1974 ruling, each state may adopt its own trade secret rules.

friend's couch. In due course, the friends then joined together as coworkers at a competing company. Such incidents leave the trade secret owner in a difficult position.

Trade Secret Law is typically statutory and generally set forth under state law. Trade secrets, such as proprietary extraction methods or unique *Cannabis* strain cultivation techniques, are invaluable assets for *Cannabis* companies. Protecting trade secrets involve maintaining strict confidentiality and implementing robust security measures.

Trade secret enforcement

This relates back to the common law "illegality doctrine" discussed above. Theft of trade secret is pursued in federal court. Because marijuana is illegal federally, state action may be preferred. There is not a lot of experience in *Cannabis* trade secret litigation, so results are inconsistent.

It is my experience that the *Cannabis* industry culture limits the efficacy of trade secret protection and so does the geographically fragmented nature of the industry. It should still be pursued where possible, but it is easier to maintain if the trade secrets are combined with other forms of IP protection.

Intellectual property (IP) licensing

IP licensing is governed by standard contract law. A few states have developed specific statutes addressing *Cannabis*-related IP licensing. Laws may require submission of confidential agreements to regulatory agency and be subject to public disclosure.

Another advantage of licensing IP is that licensing royalties paid can reduce the cost of goods sold under Section 280E. This reduces the tax liability for income to a company handling *Cannabis*. For example, if an inventor invents a product or process, they can license it to a *Cannabis* company. If the license is structured properly, the *Cannabis* company may be able to deduct the cost of goods sold, including the royalty for particular products. There is not a prohibition for an inventor to license his/her patent to his/her own *Cannabis* company. One would need to hire a *Cannabis*-savvy CPA or tax attorney to develop any compliance plan in dealing with Section 280E issues.

Pharmaceutical companies have paved the path to patentability for pharmaceutical inventions for over a century. Under patent law there is no requirement for legality. In fact, most pharmaceutical products and processing methods are not legal when a patent is filed. These pharmaceutical products are only legal after extensive clinical trials and FDA approval, which can take years and even a decade to complete. This means that the patent office grants patents to applicants for substances that may, or may not, have been FDA approved. Applicants for *Cannabis* patents thus face no legality hurdle.

Historical note about trademarks

Trademark law was the first consumer protection law in the United States. It evolved in England many hundreds of years ago. When a beer barrel cart came through a village, it often included the beer maker's mark on the barrel. Customers would recognize this as a mark of origin and, thus, quality. Some less scrupulous brewers copied the mark and sold low-quality beer to the village, making the village sick with poor quality beer. One guess is that the unscrupulous vendor used moldy grains. The king stepped in and introduced the earliest trademark laws to protect the villagers from such deceptive sales.

In the United States, the common law, which is based on early English jurisprudence, made actionable the mislabeling of products. The common law still applies in most US states. Later in time, state statutes codified the principals of common law. Both are enforceable.

Federal trademark law was introduced and codified as interstate commerce grew. Now both state and federal Trademarks can be registered. A trademark owner can choose to enforce its marks using the common law, state law, or federal law as applicable.

Federal trademarks

Federal trademark registration with the USPTO remains a significant challenge for *Cannabis*-related businesses due to the ongoing federal prohibition of marijuana as a Schedule I controlled substance, making it ineligible for federal trademark protection. Products containing isolated CBD, a hemp-derived cannabinoid, may be ineligible for federal trademark protection because the FDA has never approved CBD as a "drug." The FDA approved Epidiolex, an oral cannabidiol solution, for the adjunctive treatment of seizures in patients aged 2 and older with Lennox-Gastaut syndrome or Dravet syndrome. This approval is limited to drugs that contain *Cannabis*-derived CBD and not more than 1% THC. This is the only formulation of CBD that has received FDA approval.

Several innovative steps to secure trademark protection

Companies that produce ancillary products or offer services related to *Cannabis*, rather than directly handling the plant, can often secure federal trademark protection. This includes businesses involved in *Cannabis* accessories, packaging, consulting, or technology solutions.

Some *Cannabis* companies have successfully registered trademarks in non-*Cannabis* -related classes of goods or services. For example, a *Cannabis*-infused beverage company may secure a trademark for its product under the class for beverages, even if *Cannabis* is one of its ingredients. A multiclass application means that the same trademark is registered in multiple categories that are related to the goods or services for which the trademark is used.

Businesses can file "intent-to-use" trademark applications for *Cannabis*-related products or services, signaling their intention to use the mark once federal regulations change. This strategy allows companies to secure their place in line for federal trademark protection.

In August 2023, according to press reports, the Department of Health and Human Services sent a recommendation to the Drug Enforcement Agency to reschedule marijuana to Schedule III. This will open the door to trademark protection for marijuana products.

State trademarks

In states where *Cannabis* is legalized for medical or adult use, businesses can obtain state-level trademark protection. State trademarks are limited in scope, protecting the brand only within that specific state. However, they offer valuable protection against infringement within the state's jurisdiction. State trademark registrations allow *Cannabis* companies to build brand recognition and help businesses establish a strong presence and build brand loyalty within their specific market. Companies can take legal action against potential infringers operating within the state, preventing brand dilution and protecting their market share.

If licensing technology across state lines, a state registration in one state by a licensor can be registered in another state by the licensee who agrees to later assign the second state registration back to the original licensor. This is a nice way for a licensor to expand its trademark portfolio and obtain trademarks across many states via a licensing strategy.

State trademark registrations are typically cost effective. When *Cannabis* is rescheduled, a plurality of state registrations may make a nice paper trail for future federal registration and possible enforcement.

Copyright protection at the federal level extends to various creative works in the *Cannabis* industry. A copyright protects expression whether written, vocalized, printed, or distributed digitally. This is wide reaching and can protect an emoji, artistic logo, product label, web pages, photos, and many other creative expressions.

Cannabis businesses can register copyrights for their branding materials, including logos, product labels, and packaging designs. Protecting these elements ensures that competitors cannot imitate the visual identity of a successful brand. Copyrights also cover advertising campaigns, including print materials, digital content, and multimedia advertisements. Creative and engaging advertising can be a significant driver of brand recognition and customer loyalty.

Many *Cannabis* companies produce educational content to inform consumers about their products and the industry. Copyright protection prevents others from using or reproducing these materials without permission.

Artists and illustrators within the *Cannabis* industry may create original artwork for various purposes, from strain labels to promotional posters. Registering copyrights for these works safeguards the artists' IP rights.

Copyright holders can explore licensing opportunities, allowing others to use their copyrighted works in exchange for royalties. This can be a lucrative revenue stream for *Cannabis* businesses with valuable copyrighted content. Copyright protection is vital for safeguarding creative works and marketing materials, providing legal recourse against unauthorized use and enhancing the overall value of a *Cannabis* brand.

As a general rule, for works created after January 1, 1978, copyright protection lasts for the life of the author plus an additional 70 years. For an anonymous work, a pseudonymous work, or a work made for hire, the copyright endures for a term of 95 years from the year of its first publication or a term of 120 years from the year of its creation, whichever expires first. If somebody infringes on your copyright, you are entitled to file a lawsuit in federal court to enforce your rights. Remedies include obtaining an injunction or restraining order to prevent additional violations, an award of money damages, and possibly attorneys' fees.

It's important to note that while states have taken the lead in regulating the *Cannabis* industry, the federal illegality of *Cannabis* continues to create challenges, including issues related to banking, interstate commerce, and IP protection.

Intellectual property

If *Cannabis* is rescheduled or descheduled in the United States, it would mark a significant shift in the legal landscape for the *Cannabis* industry resulting in innovations and opportunities:

- A surge in research and development activities in the *Cannabis* industry. This could result in the creation of new and innovative *Cannabis*-related products, technologies, and processes. Patents would be crucial for protecting these innovations, brand identities, and incentivizing further research.
- The development of pharmaceutical-grade *Cannabis* products. Pharmaceutical companies would be more inclined to invest in *Cannabis* research, and patents would play a pivotal role in protecting their proprietary formulations, delivery methods, and treatment applications. There may be tension between existing products and *Cannabis* entrepreneurs and large pharmaceutical interests.
- Encourage increased competition in the *Cannabis* industry. Patents would enable companies to differentiate themselves from competitors by securing exclusive rights to their unique *Cannabis* strains, cultivation methods, extraction processes, and product formulations.
- U.S.-based *Cannabis* companies might seek to expand into international markets where *Cannabis* regulations are evolving. Patents would help protect their IP rights abroad, enabling them to establish a strong foothold in emerging global markets.
- Patented *Cannabis* strains and products could become powerful brands in the industry and extend geographic reach with ability to engage in interstate commerce. Businesses could build consumer trust by offering consistent, high-quality products with unique characteristics protected by patents. This could foster brand loyalty and customer recognition.
- Companies with valuable *Cannabis* patents could enter into licensing agreements or partnerships with other industry players, generating additional revenue streams. Licensing patented technologies or strains to other businesses could become a common practice in the *Cannabis* industry.
- Patents provide a legal framework for enforcing IP rights. Businesses could more effectively protect their patented innovations through litigation, preventing unauthorized use and infringement.
- Innovations in extraction techniques, refining processes, and product formulations would become patentable. This could lead to a more diversified and sophisticated range of *Cannabis*-derived products on the market.
- The availability of strong patent portfolios could make *Cannabis* companies more attractive to investors. Investors may place higher value on businesses with a competitive edge and significant IP assets.

It's important to note that the specific impact of patenting in a rescheduled or descheduled *Cannabis* industry would depend on the regulatory framework put in place at the federal and state levels. Additionally, any changes in *Cannabis* policy should consider the need to balance IP protections with promoting competition and ensuring access to safe and affordable *Cannabis* products for patients and consumers. If *Cannabis* were to be rescheduled in the United States, it would signify a significant shift in the legal status of the plant, potentially opening up a host of opportunities for entrepreneurs and the industry. In this scenario, patents would become even more crucial for *Cannabis* businesses.

With a more permissive regulatory environment, *Cannabis* companies may be more inclined to engage in partnerships and cooperative agreements. Patents would play a pivotal role in such arrangements, enabling companies to monetize their IP.

As the industry grows, so does the potential for disputes over IP. Patents would provide a legal framework for businesses to defend their innovations and challenge potential infringers. Patents often require a high level of detail in their descriptions. This can contribute to the development of industry standards and best practices, driving overall quality and innovation in the *Cannabis* sector.

Conclusion

Whether *Cannabis* is rescheduled in the United States or not, IP will become a cornerstone of the industry. Patents will always serve as a vital tool for protecting innovations, differentiating businesses in a competitive market, and facilitating collaborations and licensing agreements.

Experts in the Field Interview: Ethan Russo, MD

For video of this interview scan here

Biography

Ethan Russo, MD, is a board-certified neurologist, psychopharmacology researcher, and author. He is the founder and CEO of credo-science.com. Previously, he was director of research and development of the *International Cannabis and Cannabinoids Institute* (ICCI) based in Prague, Czech Republic. Medical director of *PHYTECS* (2015–2017), a biotechnology company researching and developing innovative approaches targeting the human endocannabinoid system, and from 2003 to 2014, he served as senior medical advisor, medical monitor, and study physician to *GW Pharmaceuticals*, United Kingdom, for numerous Phase I–III clinical trials of Sativex® for alleviation of cancer pain unresponsive to optimized opioid treatment and initial studies of Epidiolex® for intractable epilepsy.

He graduated from the University of Pennsylvania (Psychology) and the University of Massachusetts Medical School, before residencies in Pediatrics in Phoenix, Arizona, and in Child and Adult Neurology at the University of Washington in Seattle. He was a clinical neurologist in Missoula, Montana, for 20 years in a practice with a strong chronic pain component. In 1995, he pursued a 3-month sabbatical doing ethnobotanical research with the Machiguenga people in *Parque Nacional del Manu*, Peru.

He has held faculty appointments in pharmaceutical sciences at the University of Montana, in medicine at the University of Washington, and as visiting professor at Chinese Academy of Sciences, Harvard University, and Johns Hopkins University.

He is a past-president of the *International Cannabinoid Research Society* and is former chairman of the *International Association for Cannabinoid Medicines*. He serves on the Scientific Advisory Board for the *American Botanical Council*. He is author of *Handbook of Psychotropic Herbs*, coeditor of *Cannabis and Cannabinoids: Pharmacology, Toxicology and Therapeutic Potential*, and author of *The Last Sorcerer: Echoes of the Rainforest*. He was founding editor of *Journal of Cannabis Therapeutics;* selections of which were published as books: *Cannabis Therapeutics in HIV/AIDS, Women and Cannabis: Medicine, Science and Sociology, Cannabis: From Pariah to Prescription,* and *Handbook of Cannabis Therapeutics: From Bench to Bedside.* He has also published numerous book chapters, and over 50 articles in neurology, pain management, *Cannabis*, and ethnobotany. His research interests have included correlations of historical uses of *Cannabis* to modern pharmacological mechanisms, phytopharmaceutical treatment of migraine and chronic pain, herbal synergy and phytocannabinoid/terpenoid, and serotonergic and vanilloid interactions.

He has consulted or lectured on these topics in 39 U.S. states and Canadian provinces and 39 countries.

Additional information can be found at *Research Gate* and *LinkedIn*.

Jahan Marcu:

Hello, I am Dr. Jahan Marcu.

Andrew Peterson:
And I'm Dr. Andrew Peterson, and we are the editors of the *Cannabis* Innovation Series, and we'd like to welcome you to yet another experts in the field interview. This volume of the series is focused on the regulated *Cannabis* and hemp market navigation, examining business, public health, and research challenges.

Jahan Marcu:

We are here today to talk about biotech, public health, and research challenges. With us today is Dr. Ethan Russo, founder, CEO of CReDO Science. He is a world-renowned clinician, scientist, educator, and entrepreneur in the *Cannabis* space.

Andrew Peterson:
So for me it's interesting to talk with someone with such experience and exposure in the *Cannabis* field. So Dr. Russo, if you wouldn't mind, tell us a little bit more about yourself, about CReDO Science, and how did your career path get you there . . . or here?

Ethan Russo:
Sure, first it's a pleasure to be with you. I'm a neurologist by training. Went through standard pharmacology training, et cetera, et cetera. However, by the time I was in practice in 1990, for seven years at that point, I really became disillusioned with the kinds of results that we were getting with conventional medicine. Neurology is a tough thing. You don't often succeed on your first trial of a medication. So it was really a revolving door of people coming back to treat the chronic pain or seizures or migraines, and it was sort of the drug of the month club.

So at that point, I really rekindled an interest that I had in medicinal plants, which, to shorten the story up, led me to the rainforest in Peru and on sabbatical in 1995. But when I came back, it was 1996, it was Prop 215 in California, leading to the revolution in medicinal access to *Cannabis*. And I quickly became involved in the system, trying to do research and finding that I was being thwarted by the government. But a new love affair began in terms of the study of *Cannabis* and the endocannabinoid system, and it basically led to a second career.

So after doing a bunch of writing, editing, consulting, in 2003 – it came on full-time for 11 years with GW Pharmaceuticals and the development programs for the pharmaceutical, *Cannabis*-based medicines, Sativex and Epidiolex. Subsequently, it was with a couple of private companies culminating in 2020 with the development right before the pandemic. So that set us back a bit, but CReDO Science is devoted to making *Cannabis* safer and better. We have seven of our own patents pending, various technologies running the gamut from a diagnostic test for cannabinoid hyperemesis syndrome to extraction techniques and development of what we hope will be newer and better formulations that are *Cannabis*-based, both for the supplement market and pharmaceutical markets.

Jahan Marcu:
Wow.

Andrew Peterson:
Wow.

Jahan Marcu:

The average *Cannabis* and hemp company in the industry doesn't seem interested in conducting research, science, publishing data, or intellectual property development, which is kind of all these things you've touched upon. But how do companies make a decision to invest in *Cannabis* research? How might that change in the future?

Ethan Russo:
Well, right now it's pretty easy. Nobody's investing. So there's a simple explanation. It may be an old figure, but as of a couple of years ago, 85% of the investment in *Cannabis*-based companies was solely in publicly traded companies. We've chosen to be private and will remain so just as a matter of control. I think part of the problem in this situation is when you're publicly trading, you are really responding to what the stockholders want, rather than what the industry might want or what science and medicine require. So we're looking at this from a very different standpoint. We feel it's the only way to go. Part of this is motivated by the fact that I'm old now. I've had a lot of lousy bosses in my life. Currently, I'm the boss of this company, and if there's a problem, I've only myself to blame. So it's better that way.

Andrew Peterson:
So, sort of related a little bit, you said that most of them are public, publicly traded and you're privately traded. What's making them publicly traded? Why do they go in that direction?

Ethan Russo:
Well, I think I . . . Basically, again, it's a philosophy I don't espouse. I understand the need to be publicly traded, but the idea seems to be that we have an idea that we can sell and we're gonna try and make a bunch of money and then sell it off. And that's what people do. But to pick on a specific company, Canopy spent $50 million acquiring other companies and very little concrete came out of it. So, some people made a bunch of money. I can't see that a lot was produced or the public health was improved as a result of all these machinations.

Andrew Peterson:
So I'm wondering a little bit then, because recently the Department of Health and Human Services recommended *Cannabis* to be placed in a schedule three. Does that affect you, or will it affect your work, and even would it affect some of the other ones, the public, public retreat?

Ethan Russo:
Uh . . . Probably not affecting me much at all. Other companies, yes, but mainly in terms of, my understanding is it would ease up some of the banking restrictions. It

would also, for some, improve access to *Cannabis* for research. However, I think that the premise of the exercise is pretty faulty. Schedule III in the Controlled Substances Act an FDA approved pharmaceutical that would have limited drug abuse liability. It really doesn't apply to *Cannabis* at all. So it's possible that . . . this would be interpreted as being restricted to drugs that were made from *Cannabis* that were already FDA approved. Right now, there's only one of those, Epidiolex, which is 97% pure CBD. Sativex, that I mentioned before, is approved in 30 countries, but extra USA. So I think if it were misapplied, it could actually make things worse for the rest of the *Cannabis* industry. Potentially, it might be possible to eliminate all the commerce and CBD products and other products that are currently available. So I just don't understand it. My personal feeling is that *Cannabis* should be descheduled and then regulated for safety and other considerations. So it might be akin to what's done alcohol but should be distinct.

Jahan Marcu:
Very interesting, you know, and I think with some of these changes we might see in regulations, as *Cannabis* moves down the scheduling pathway . . . I don't think schedule III will be its final destination if indeed it does end up there you know, I'm thinking about as the industry matures we might see different partnerships. We've already seen companies start to partner with universities. We've seen beverage companies partner with *Cannabis* companies. We've seen ingredient suppliers . . . Now, what I wanted to ask you, is to peer into your crystal ball and talk about what might a business partnership or relationship between a *Cannabis* company and a pharmaceutical company look like? I mean, would the pharmaceutical company buy the rights to all their products, just like they buy portfolios of small biotech or . . . Would it be more like a Rite Aid-type over the counter model? I don't know . . . we'd be curious to get your perspective. What do you think about that?

Ethan Russo:
Sure. Well, any of these things might be possible. You know, a good model would be that from Canada, which is a lot more supportive of research in general on *Cannabis* research in particular. They have a program, as one nice example, called Mitacs, in which a graduate student or a postdoctoral fellow can get half of their salary supplied by the Canadian government, with the other 50% supplied by the private company running the research. We don't have anything like that in the United States. Basically, everything is funneled through the National Institutes of Health, which has really been strangled politically. I won't say which party, but people can likely infer. Additionally, certainly, it is the case over the last four decades that the vast, vast majority of research in *Cannabis* has been devoted to its harms rather than its pharmaceutical benefits. So that remains a problem. Additionally, it's very hard to work with American universities because of their insistence or requirement to impose vast overhead expenses. Just as a comparison, I can tell you the GW pharmaceuticals for whom I worked didn't have these restrictions in other countries. So they had many, many collaborators, particularly in European countries, that would get funding from them and do the research without 60% or whatever the figure would be being siphoned off by

the University for Overhead. So basically none of the research that GW underwrote that really told us most of what we know about cannabidiol, cannabigerol, etc. came from this underwriting by a private company, but almost none of it was done in the United States.

Jahan Marcu:
Wow, so it seems like incentivizing these partnerships will be key because doing it through traditional pathways where 60% of the overhead is taken, in an industry where you still can't take tax write-offs, just may not be sustainable.

Ethan Russo:
No, I think it's damn near impossible.

Andrew Peterson:
But that's where if you reschedule to a three, the tax benefits, the 280e doesn't apply anymore, so that might be a benefit. But still, 60% of the total grant money going to some universities is an exorbitant amount.

Ethan Russo:
Sure.

Andrew Peterson:
I want to kind of change this a little bit and think a bit more about, because you mentioned some of this, about some of the harms that occur and some of the benefits that occur. How should *Cannabis* businesses balance, like consumer safety versus sales? I mean, right now they're struggling because they need to make sales to be able to make some money, but then consumer safety becomes an issue. How do they balance that now? And then how do you see that coming out in the future?

Ethan Russo:
Well, it all depends on how the rules, that are made by politicians, come through in the situation. I don't think that it is advisable or possible to shortcut safety. That's got to be number one, along with quality control. So we're just not willing to compromise on that score. Now, obviously, it adds to the cost if you're requiring a certificate of analysis on anything that is sold to a consumer or that you're requiring pesticide screening and screening for heavy metals and all this. It adds up, but it's a situation where no one in good conscience should be selling product to consumers without knowing that it's going to be safe – when used as directed.

Obviously, there can be limiting factors if people get too much THC. Most of the other components of *Cannabis* are extremely low in toxicity and risk of adverse events. If it's not made properly and is contaminated in some fashion, it wouldn't matter how good the base ingredients were. So there can't be any compromise on safety. And whatever that ends up costing, the consumer is going to need to bear. Hopefully, their return in terms of the benefit that they receive will make that worthwhile.

Jahan Marcu:
I really like what you said about there's no compromise on safety. And, you know, some people in the *Cannabis* space, I think, especially younger professionals getting into space are looking at the industry and kind of scratching their heads because one company seems to be really into research and applying data and others could care less. And so I have been seeing more young. Entrepreneurs going into biotech because they can, as you said, raise that private capital, avoid the costs of doing research with the academic collaborator. But we're still in a period of change, like a rapid change where things change every week in the *Cannabis* space, it seems like. So I was wondering if you could just opine a little bit on advice for young biotech entrepreneurs, looking at, maybe, going the similar GW Pharma route that you went, or perhaps trying to develop their own cannabinoid product, whether a natural product or synthetic product, and taking it to traditional routes for federal approval.

Ethan Russo:
Sure. Well, you know, it's a situation that . . . there are always going to be three echelons of activity with respect to *Cannabis*. One is *Cannabis* herb, which usually is going to be in the form of flower. Then there are the more refined products that would include your CBD products and oils that presumably will continue to be sold over the counter. And finally, the top echelon would be pharmaceutical materials, that go through regulatory process with the FDA and other bodies around the world.

I think people make a mistake when they think that one aspect of the market can eliminate the others. They're always going to be there and they need to coexist. We're working in all three areas, and that is because we think that they all have their place and role towards the public health.

So in terms of advice, I think that there are two ways to gain experience. One is through academic study, and the other one is on the job. It remains the case that there are very few universities with programs. I had mentioned before we began that I lectured at Humboldt in California on Friday. They've developed an undergraduate major in *Cannabis* studies, right now it's sociologically based. I think that they would like to expand that into more technical aspects, but they're not currently allowed to grow. If there were *Cannabis* legalization, they could have a botany program and more commercial aspects. There also is a master's program in Maryland at the university.

So these are really good opportunities. Otherwise, people are learning on the job, usually, you know, under black market conditions or at least until recently in most places. Again, there are a lot more opportunities abroad. I mean, my life is a personal example. I've spent most of the last 27 years abroad in which I've been engaged in *Cannabis* working for foreign companies. And that was through necessity. I just couldn't do the kind of work that I felt was required in the United States.

Jahan Marcu:
That is really fascinating stuff. And I know that you've answered a lot of questions from young professionals about what they should do and how they should engage in this. But I like what you said is that there's a lot of education and certification opportunities out there. There's a lot of ways to get experience and not don't be afraid to go abroad. Don't be afraid to get experience that might help you with your, with your pathway forward.

So, that brings me to a follow-up question, you have a lot of publications out there, a prolific writer about *Cannabis*. I was wondering if there is a resource that you find yourself sharing most often with other curious *Cannabis* minds? Is there a book? Is there a podcast episode? Maybe it's something else that you find yourself sharing most often when people say, "hey, what do you have about *Cannabis* that I should read or I should listen to?"

Ethan Russo:
Well. There are a number of good books at this point by Dustin Sulek, Bonnie Goldstein, my business partner, Nisha Whiteley. All of these are available online. In terms of where I point people the most often, it's actually my personal website, heathenrusso.org at the library tab. published peer-reviewed articles and book chapters there. Because it's my personal site, there are no copyright issues and people can download and print those as they wish. So if people are interested in the work that I've done, I would send them there first. Because again, it amounts to a lot of paper available. So yeah, I hope people will avail themselves of that opportunity and I appreciate the question.

Andrew Peterson:
Well, I think I'm going to go there and check out what you have. I didn't realize you had that library of resources, so I do appreciate that. I think a lot of people will after they hear about it from this and other sources. So, thank you. So for us, Dr. Russo, we are at the end. We've finished the questions that we are interested in. Anything final, any final comments you want to say for the audience?

Ethan Russo:
No, again, I appreciate this opportunity. I like the work that you're doing and hope that you'll be very successful in reaching a wide audience.

Jahan Marcu:
Thank you so much.

Andrew Peterson:
Thank you for coming out. We do appreciate your time.

Miles W. Tuttle

The U.S. *Cannabis* experiment: navigating state-specific policies, ethics, and standards in the absence of interstate commerce

A single courageous State may, if its citizens choose, serve as a laboratory; and try novel social and economic experiments without risk to the rest of the country.
– U.S. Supreme Court Justice Louis Brandeis

Introduction

An experiment is a scientific procedure undertaken to make a discovery, test a hypothesis, or demonstrate a known fact. There are currently over three dozen individual states serving as laboratories, courageously experimenting with novel, autonomous social and economic *Cannabis* policies, ethics, and standards, utilizing its "citizens" as willing subjects in their experimentation; however, the state-circumscribed isolation of current business models imposed by the prohibition of interstate commerce creates inconsistencies among business practices, resulting in financial and operational inefficiencies, social costs, public health risks, as well as other potential unintended consequences.

From the Controlled Substances Act of 1970 to the Farm Bill of 2018, this swinging pendulum of legality has had a direct and substantial effect on *Cannabis* industry business practices, ranging from banking and taxes to interstate commerce and counterfeit products. While increases in accessibility and availability of "legal" *Cannabis* products could be perceived as a potential advance in business opportunities, it could also lead to market oversaturation, price compression, and industry layoffs. Following the pandemic stimulus-fueled record sales of 2020, consumers' "willingness to pay" has decreased sharply, forcing market corrections in pricing and human resources [1]. As this "gray" legal status of *Cannabis* continues to shift, increased availability and accessibility could continue to decrease the scarcity, value, and risk associated causing this "gray" legal window of opportunity to close quickly, squeezing out both players and profits along the way.

As varying ethical positions have shifted over time, moral principles and business ethics have similarly adjusted from the racial disparity of the "war on drugs" and "mass incarceration" to the attempted inclusivity of recent "social equity" and "prisoner expungement" good faith efforts. Once considered negative to associate a *Cannabis* business with minorities or prior drug convicts, states are now encouraging and even requiring the inclusion of these alleged, former "black" market individuals previously

Miles W. Tuttle, University of Hawaii, Richardson School of Law

https://doi.org/10.1515/9783110677508-007

harmed by the perceived poor ethical judgment of the past. While morals represent guiding principles, ethics reflect behaviors or actions implemented; however, based on current minority *Cannabis* market participation statistics, it may appear that these varying state-specific ethical actions are nothing more than a good faith effort.

As this patchwork of individual state-regulated *Cannabis* programs experiment with solutions and standards, mature, established industries (such as pharmaceutical, alcohol, tobacco, and consumer packaged goods) have positioned significant investment capital into potentially incorporating *Cannabis* into their product lines possessing preexisting standards. With billions of dollars to invest and decades of industry knowledge, experience, and successful operation, these seasoned industries can easily modify established standards to accommodate the necessary requirements of *Cannabis*-based products. While these battle-tested industries with extensive national distribution channels have previously overcome struggles with their own public health risks and issues, drawing from this knowledge and experience could directly translate to mitigating parallel public health risks and issues associated with *Cannabis*. As each of these individual burgeoning states make their own attempts to preinvent the wheel, proven preexisting standards have already been validated by other industries.

An experiment is a scientific procedure undertaken to make a discovery, test a hypothesis, or demonstrate a known fact. As over three dozen individually sequestered laboratories continue to perform their own experiments, the results could continue to reflect a lack of uniformity and standardization as one of the major obstacles to overcome. As with oranges commonly grown in Florida, or cheeses popularly produced in Wisconsin, neither of these industries may perform as well when circumscribed within the borders of Alaska. Likewise, constraining *Cannabis* policies and products within the isolated borders of each state could produce a dearth of product diversity, inconsistent consumer education, and assorted state-specific health and safety risks. As an industry projected to attain $52.6 Billion by the year 2026 (Figure 1), the potential influence of permitted interstate commerce and a singular uniform concept of acceptable policies, ethics, and standards could lead to the discovery of possible consolidation and streamlining of operations, achievement of stable price equilibrium across all states, and specialized production capitalizing on regional advantages [2].

Price compression

During the gold rush it's a good time to be in the pick and shovel business.

– Mark Twain

Once a product worth its weight in gold, recent stumbles have forced drastic changes and pivots to accommodate the plummeting perceived value of this once scarce, prohibited commodity. Isolated by the prohibition of interstate commerce, multistate opera-

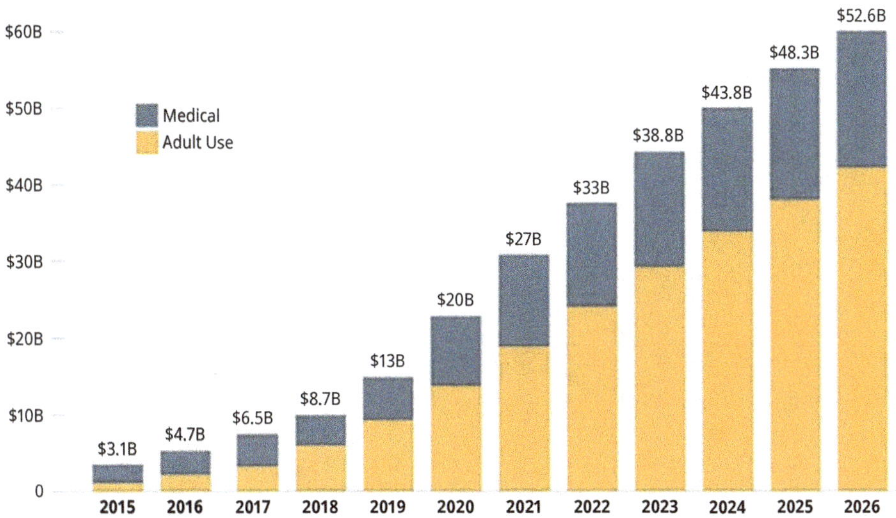

Figure 1: U.S. *Cannabis* Retail Sales Estimates: 2015–2026 [2].
Source: MJBiz Factbook.
@2022 MJBiz, a division of Emerald X, LLC.

tors have been siloed within the policies of varying state borders. As new entrants, both legal and illicit, participate in these oversaturated markets, multiple *Cannabis* companies are abandoning established, mature states in search of greener pastures. While *Cannabis* market policies evolve in legalized states and regions, product saturation and price compression could stymie profitability and growth opportunities. Although individual, state-specific *Cannabis* policies continue to enable increased "legal" accessibility and availability, mature markets operating for over five years are experiencing the effects of declining prices and diminishing profits (see Figure 2).

Cannabis industry business practices have undergone significant changes over the years, shaped by various policies and regulations. Historically, *Cannabis* was illegal in the United States under the Controlled Substances Act of 1970, which classified marijuana as a Schedule I drug [4]. This prohibition made it legally impossible for businesses to operate in the *Cannabis* industry at that time, leading to the emergence of the "black" (or illicit) market. In 2022, BDSA[1] Market Forecast estimated that the illicit component made up roughly 55% of total *Cannabis* dollar sales across U.S. markets with active adult-use or medical programs [5]. Currently there are varying levels of legality and subsequently varying sizes of illicit markets in each of the 50 "United" States.

1 BDSA, previously recognized as BDS Analytics, is a hub of expertise and research on the present and future of the legal *Cannabis* market specializing in gathering, analyzing, and presenting data on point-of-sale market share and consumer surveys for companies within or influenced by the *Cannabis* sector.

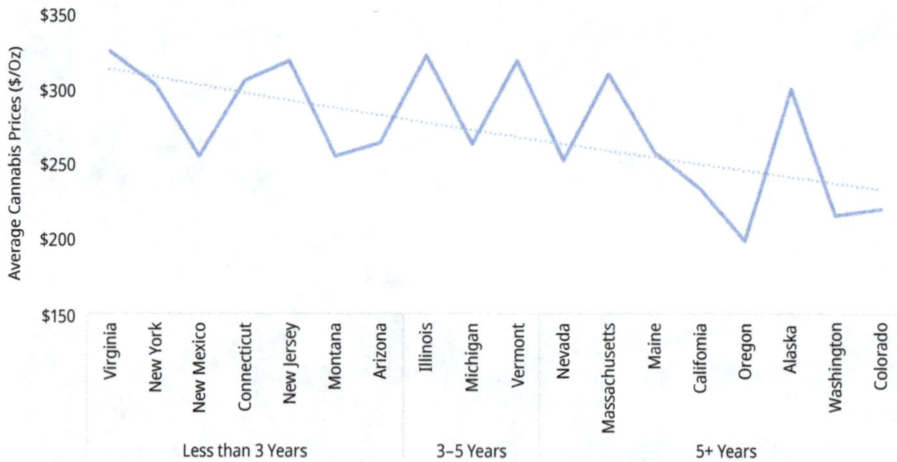

Figure 2: Fully Legalized States: Age of Legalization vs. Average *Cannabis* Prices ($/Oz) [3].

While some states have legalized *Cannabis* for medical and/or recreational use, its current federal Schedule I listing denies businesses the accommodation of transporting *Cannabis* across state borders [6]. As a result of these restrictions on interstate commerce, many *Cannabis* businesses are limited to operating within their own state boundaries and adhering to their own set of regulations and policies, which can limit their potential customer base and reduce their ability to achieve economies of scale [6]. Mature markets, such as Oregon and Colorado, experienced a rapid increase in the number of licensed cultivators and dispensaries due to *Cannabis* legalization policy change, leading to an oversupply of products and declining prices [7]. According to a recent study, the average price of *Cannabis* flower in Colorado declined by over 60% between 2021 and 2022 [8]. With 25,000–33,000 *Cannabis* businesses in the United States in 2019 alone [9], market oversaturation and interstate commerce restrictions have made it challenging not only for new entrants to gain a foothold but also for established businesses to stand out in the market, resulting in increased competition, reduced profitability, and ultimately demand for dramatic pivot. This regulatory and economic burden has resulted in the streamlining and/or relocating of large *Cannabis* companies to nascent locales, as well as several *Cannabis* business closures, particularly seen in small-scale operators that relied on higher prices to offset their modest scale of operations [10].

Take for instance the miner who discovers a profitable, untapped prospect to pan during the Gold Rush. As more miners proliferate this prospect, his pan becomes less profitable, until he is forced to leave and discover a new untapped prospect. Replace the miner with a fisherman on a pond, and a similar concept of market oversaturation applies. Both examples parallel the story of market oversaturation in matured *Cannabis* states; however, based on product scarcity, gold possesses a greater ability to preserve its perceived value, as unlike *Cannabis*, gold doesn't grow on trees.

Colorado's legal *Cannabis* market has experienced a significant decline in prices, with the average price per pound of *Cannabis* falling from $1721 in January 2021 to $658 in October 2022 [8]. California, Oregon, Washington, Nevada, and Alaska are all suffering the consequences of price compression, demonstrating that commoditization of *Cannabis* is not only apparent among individual states but regionally as well. Correspondingly, according to a recent report by Headset, the average price of *Cannabis* flower in the United States has decreased by 29% in the past year alone, due in large part to price compression [11]. While profits and prices may be waning, customer demand has remained steady or increased, revealing a decline in consumers' willingness to pay. As ticket prices are dropping, the quantity of items per ticket are not (Figures 3 and 4).

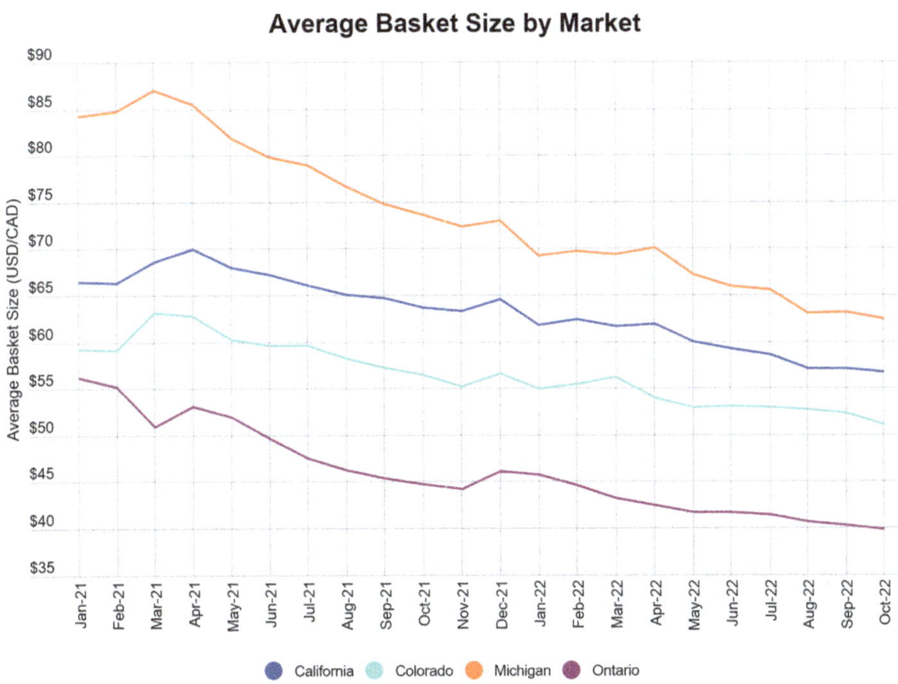

Figure 3: Average Basket Size by Market [11].

To survive in this challenging business environment, *Cannabis* companies have had to adapt their business practices. According to a report by BDSA, businesses are increasingly adopting innovative pricing strategies, such as dynamic pricing, to remain competitive in an increasingly crowded market [12]. Regulatory changes in California, such as increased taxes and regulations, have resulted in consolidation and cost-cutting measures to reduce expenses and improve margins [10]. Layoffs have become increasingly common, as businesses such as Curaleaf, Trulieve, Leaflink, and Columbia Care seek to reduce labor costs [1]. Others have focused on product differentiation,

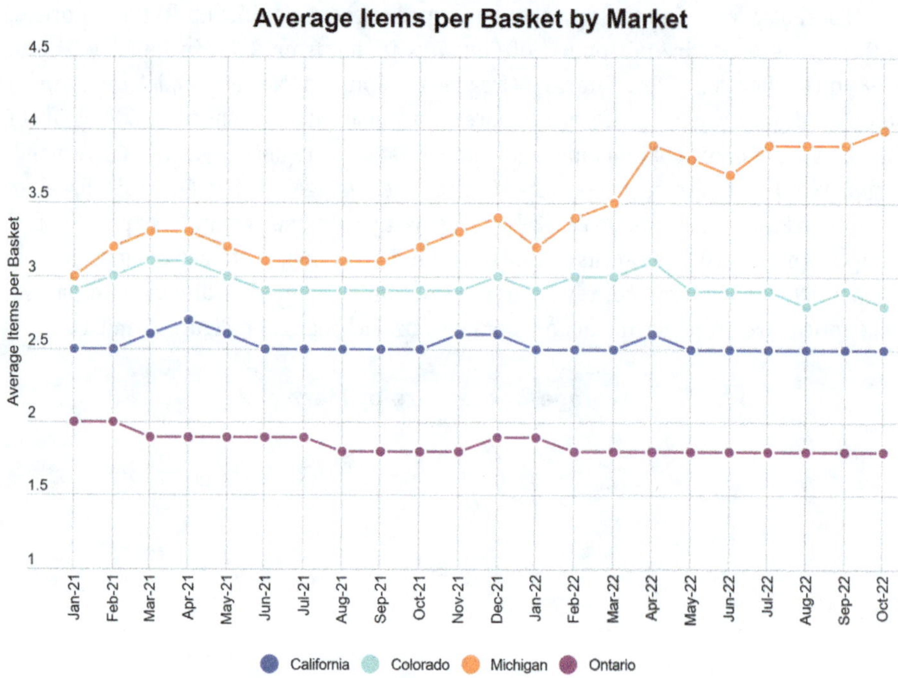

Figure 4: Average Items per Basket by Market [11].

developing new and unique products to stand out on the shelf [10]. Researchers found that many of these struggling growers are turning to alternative business models to stay afloat, including focusing on niche products and services, such as high-quality strains or personalized customer experiences [13]. Furthermore, essential ancillary products that support *Cannabis* businesses could offer a more stable revenue stream as the plant itself loses value progressively by state.

Notably in both cases of the miner and the fisherman, the suppliers of picks and shovels, as well as tackle and bait, experienced no price compression, only increased volume and revenues. Alternatively, just as the miner and fisherman were forced to move, numerous multistate operators are each implementing a similar strategic mass exodus to relocate or reallocate to newer, untapped markets.

Social equity

We knew we couldn't make it illegal to be either against the war or blacks, but by getting the public to associate the hippies with marijuana and blacks with heroin, and then criminalizing both

> *heavily, we could disrupt those communities. We could arrest their leaders, raid their homes, break up their meetings, and vilify them night after night on the evening news.*
> *Did we know we were lying about the drugs? Of course, we did.*
> – John Ehrlichman, White House Counsel under President Richard Nixon

While the Vietnam War has long since ended, the remnant criminal stigmatization of minorities associated with *Cannabis*, instigated by "shady" political strategies over 50 years ago, continues to dictate the ethical, or nonethical, business practices within today's *Cannabis* industry. As the United States has been known to embrace unethical standards in the past, each state attempts to make its own appropriate adjustments within differing timeframes and manners. In the current *Cannabis* industry landscape, each state has adopted differing stances pertaining to the legal or "white" market vs the illicit or "black" market, as well as the skin color of those permitted to participate in the industry. As many social justice and equality improvement measures have taken varying periods of time dependent upon state or region, *Cannabis* social equity programs experience similar inconsistencies across state borders. While some states make concerted attempts to require inclusivity, other states merely implement futile good faith efforts to preserve an ethical public image with no actual plan or framework to achieve a socially equitable business environment. Without consistent, "black and white" laws and universally accepted ethical business practices, the *Cannabis* industry could continue to struggle with uneven diversity and selective inclusion resulting in 50 varying shades of "gray".

The historical criminalization of *Cannabis* has been a major driver of racial inequality in the United States. "President Richard Nixon's War on Drugs wasn't really about drugs at all but was an intentional effort to criminalize Black people and disrupt their communities" [14]. According to the American Civil Liberties Union, Black Americans are 3.64 times more likely than White Americans to be arrested for marijuana possession, despite similar usage rates [15]. This racial disparity in *Cannabis* arrests has been linked to the disproportionate enforcement of drug laws in communities of color and the resulting collateral consequences for engaging in activities that are now legal. "Nothing has contributed more to the systematic mass incarceration of people of color in the United States than the War on Drugs" [16]. Furthermore, the lack of social equity in the *Cannabis* industry has also led to a paucity of investment in these communities, creating an unequal playing field for aspiring *Cannabis* entrepreneurs [17]. Beyond creating unequal access to the industry, the increasing complexity of state-specific social equity requirements and qualifications has caused businesses to have to navigate a difficult landscape to remain compliant. These businesses are unable to access the same resources and opportunities across state lines and are unable to benefit from economies of scale, thus limiting their potential growth and development.

The general purpose of social equity policy is to ensure that all members of society have equal access to resources, services, and job opportunities. However, a recent report done by Leafly found that even with social equity policies in place, only 2% of

Cannabis businesses are owned by people of color [18]. Broadly, social equity in the *Cannabis* industry falls into three core policy categories: criminal justice reforms, including automatic expungement of past *Cannabis* offenses; reinvesting a percentage of *Cannabis* tax revenue into the most impacted communities; and creating a cohesive *Cannabis* industry licensing framework with special considerations for people affected by the "war on drugs" [19]. It has been noted that the failure to implement equity in the *Cannabis* industry is largely due to the lack of resources, support, and access to capital that many minority communities face [20]. Moreover, research conducted in a 2022 annual report revealed that people of color are significantly underrepresented in executive positions, holding only 12.1% in the *Cannabis* industry [21].

In addition to this lack of minority ownership, a majority of licenses issued in the industry are given to large corporate entities, or those with higher levels of economic and social capital, leaving many minority communities unable to acquire the resources needed to participate [22]. Lofty start-up costs present a significant barrier for those who are unable to obtain traditional banking and financing services. Furthermore, this lack of access could be exacerbated by a lack of understanding of the *Cannabis* laws and regulations [17].

Compounding issues of attaining capital, Massachusetts State Representative Daniel Donahue (D-Worcester), a member of the Legislature's *Cannabis* Committee, says entrepreneurs are at risk of falling victim to "predatory loans or capital schemes" while seeking funds to support their participation in the *Cannabis* industry [23]. In Massachusetts, social equity applicants were getting "squeezed out" by exorbitant fees when municipalities were able to require *Cannabis* businesses to pay percentage-based annual Community Impact Fees and other "donations" or incentives to influence their Host Community Agreement selection [23].

In response to the failure of *Cannabis* regulation providing adequate access to social equity applicants, many states including California, Illinois, Massachusetts, and New York have begun to introduce or consider legislation to address this issue. Certain bills have introduced a tailored system of social equity licensing for *Cannabis* businesses, some mandating a minimum number of *Cannabis* licenses be awarded to social equity applicants, while others create funds for reinvestment into communities that have been disproportionately impacted by the "war on drugs".

Despite the introduction of these novel social equity policies into the *Cannabis* industry, "so far, no program has successfully achieved its social equity goals as envisioned by the people [they] serve" [19]. The failure of implementing sound, ethical social equity policies in the *Cannabis* industry thus far has had a deep and lasting impact on communities of color. As more states begin to recognize the importance of addressing social equity issues in the *Cannabis* space, a number of pending bills and other legislation are and will be proposed with the *intention* to help level the playing field, establish clearer, uniform, "black and white" laws, and ensure that the *Cannabis* industry is more equitable and accessible to all.

Public health risks

> *It was a menacing marketplace filled with products . . . [that] were, at minimum, useless remedies that picked the pocket of the user, but they could also be downright harmful.* [24]
>
> – John Swann, PhD, historian at the Food and Drug Administration

At the turn of the twentieth century, snake oil salesmen could freely peddle products asserting false health benefits and unsubstantiated safety claims. With the introduction of the Federal Food, Drug, and Cosmetic Act of 1938, along with its 1962 Kefauver Harris Drug Amendments, manufacturers were required to produce proof of safety and efficacy prior to marketing [24]. Nearly a century later, the U.S. *Cannabis* industry is confronting a similar issue based on the absence of uniform national standardization. A recent proliferation of laboratory testing result inaccuracies and misconduct has produced negative implications on *Cannabis* businesses. These scandals have adversely impacted brand reputation, resulting in loss of business, fines, legal action, the revocation of licenses, as well as the erosion of consumer confidence and its subsequent negative financial impact and potential public health and safety risks [25]. Despite state-level standardization paving the way for increased quality control within all aspects of the *Cannabis* industry, the lack of a uniform national standardization system has led to an ambiguous patchwork of regulation resulting in decentralized decision-making and considerable risk taken on by *Cannabis* businesses and consumers alike.

Standardization is a key component to ensuring that *Cannabis* products are safe for consumption. Without standardization, there is no guarantee that a product contains the amount of active ingredients that are stated on the label and the quality of the product is at risk of being inconsistent, compromised, or improperly labeled and packaged. This negligence could lead to an array of potential public health risks, such as the accidental ingestion by children, inaccurate dosing, and the risk of consuming a product that is contaminated with hazardous chemicals. Additionally, without standardization, it is difficult for businesses to accurately track and monitor the quality and safety of their products, leading to increased costs and decreased efficiency.

Even following the introduction of state-regulated *Cannabis* packaging polices, adverse *Cannabis* exposures increased nationwide from 2,951 cases in 2016 to 11,569 cases in 2020 [26]. This overall 292% increase over five years reveals an ineffectiveness of state-regulated *Cannabis* standardization policy efforts, compared to increasing access in legalized states. Over the same five-year time span, nearly half the *Cannabis* edible exposures reported in poison control cases pertained to children under the age of 12 years. Although the mortality rate is zero, poison control cases pertaining to *Cannabis* edible exposures in children under 5 years of age increased from 7 cases in 2016 to 2,119 cases in 2020 (see Figure 5).

Additionally, *Cannabis* products have been linked to the outbreak of vaping-related lung illnesses in the United States, highlighting the need for standardization of *Cannabis* extracts to ensure safety. In 2019, Denice Grady investigated the political environment

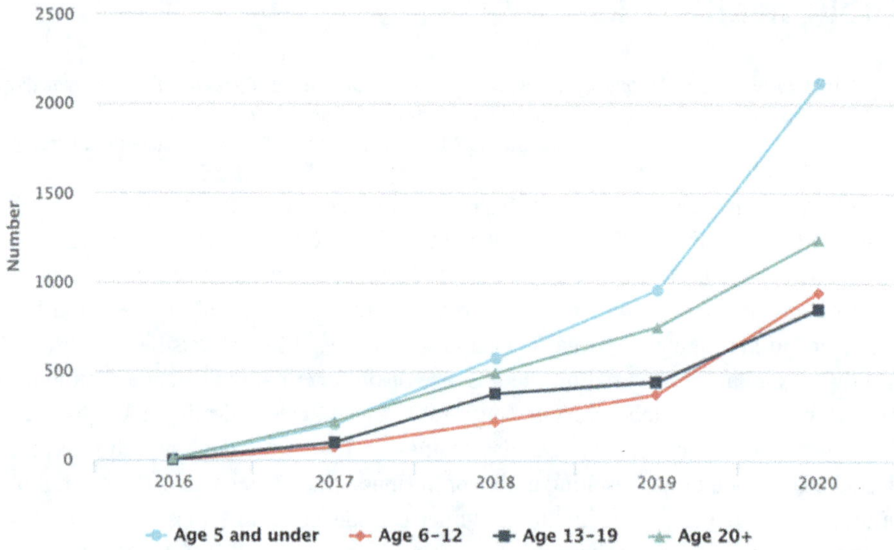

Figure 5: Marijuana Edibles Exposure Cases Reported to Poison Control Center by Age, the United States [26]. Source: NPDS.

surrounding the exposure of unethical practices regarding over 1,000 illnesses related to counterfeit branded packaging of *Cannabis* vaping products pertaining to the 2019 EVALI crisis. Grady explained that these "brands" were not a specific formulation or a single product, but rather "branded" counterfeit packaging and labeling that THC sellers could slap on any untested, homemade product [27]. The Federal Centers for Disease Control and Prevention stated that "Dank Vapes appears to be the most prominent in a class of largely counterfeit brands, with common packaging that is easily available online and that is used by distributors to market THC-containing cartridges with no obvious centralized production or distribution" [27]. As no federal regulatory body governs this industry, it is widely acknowledged that the illicit vaping marketplace is composed of a wide array of products, ingredients, packaging, and supply chains, leaving consumers with scant information on the components of the liquids they are vaping.

In an effort to address the lack of standardization in the *Cannabis* industry, many states have introduced legislation that would require businesses to adhere to certain standards for product labeling, testing, and safety. The OLCC, or Oregon Liquor Control Commission [28] as well as the Bureau of *Cannabis* Control (BCC) in California [29], have established a range of regulations and standards for their industries, including testing for contaminants, labeling specifications, and packaging requirements.

Despite these measures, the lack of a uniform, national standardization system has hindered the effectiveness of state-specific regulations. Allowing state self-regulation in the *Cannabis* industry has enabled consumers to purchase large quantities of higher potency, highly diversified *Cannabis*-based products in certain states, while studies are simultaneously showing a rise in *Cannabis*-related emergency department visits [30]. Due to the absence of consistency among states in terms of potency, packaging and labeling, and quality control of *Cannabis* products, the public is left in confusion pertaining to deciphering comparable product preferences and vulnerable to the effects of unanticipated outcomes.

Cannabis businesses trying to compete in this free-for-all atmosphere without being held accountable to national standardization are being influenced by such industry pressures as consumer demand for high THC, no federal oversight, and a flooded marketplace leading to the proliferation of "less-than-ethical" practices, such as lab shopping or choosing labs that give favorable results pertaining to potency or contaminants [31]. The 2018 Farm Bill has also created room for unethical loopholes leading to the emergence of potentially dangerous transfigurations and synthetic *Cannabis* concoctions such as Delta-10, and THC-O acetate [32].

With the *Cannabis* industry poised to continue significant growth, established monoliths such as pharmaceutical, alcohol, tobacco, and consumer packaged goods have seen the potential to incorporate *Cannabis* into their product lines and leverage their existing infrastructure, distribution channels, and experience to capture a share of the market. These prodigious industries have significant experience in navigating complex regulatory environments, including managing public health risks and issues, adhering to rigorous quality control standards, and investing significantly into the research and development of their specialized enterprises. By leveraging the experience of these seasoned industries and their lessons learned, the *Cannabis* industry could establish best practices for managing public health risks and issues associated with *Cannabis* consumption.

Without federal *Cannabis* regulation in place, there are no universal directives besides "the uncertainty that comes from mostly unenforced federal prohibition. For businesses, this creates a chaotic and unpredictable environment at both the state and local levels, which comes on top of the arduous task of securing capital and banking for a product still federally illegal" [33]. To effectively reduce the potential adverse health and safety risks of *Cannabis* use, it is necessary to develop a uniform, national standardization system that takes into account the unique needs and regulations of each state. Established industries may offer a fast track to this need for national homogeneity. For now, while individual states construct their own independent *Cannabis* product, packaging, labeling, and operational rules and regulations, the enforcement of *Cannabis* business standards will remain unchecked until the barrier on federal oversight falls with a change in national *Cannabis* policy.

The experiment continues . . .

> *Allowing states the option of moving forward now, towards a more economically rational, environmentally sustainable, and broadly beneficial supply chain will not preclude – and is in fact likely to enhance – the opportunity to fight for a smarter and more equitable federal legalization when the time comes.* [34]
>
> – Adam Smith, Alliance for Sensible Markets

As the U.S. *Cannabis* Experiment continues, many discoveries have been made and many hypotheses proven or disproven, but most facts pertaining to best practices among *Cannabis* business policies, ethics, and standards are still yet to be agreed upon. While over three dozen individual states serve as "laboratories of democracy," the continued isolation of current business models imposed by the prohibition of interstate commerce creates continued inconsistencies among business practices, resulting in financial and operational inefficiencies, social costs, public health risks, as well as other potential unintended consequences. While the authorization to transport *Cannabis* and *Cannabis*-based products across state borders could be considered positive or negative, dependent upon perspective, this inability has potentially enabled state and regionally specific price compression, varying shades of ethical inequality, and collateral health risks associated with inconsistent standardization. While the federal legalization of *Cannabis* could generate numerous opportunities and improved efficiencies, it could also exacerbate the known and unknown facts associated with this ongoing experiment. Alternatively, the implementation of state-regulated interstate commerce prior to federal legalization could lead to streamlined operations and improved profits, greater ethical inclusivity baked into the industry, and improved consistencies and standardization leading to fewer negative public health outcomes associated. These courageous "laboratories of democracy" could potentially provide the best models for federal legalization, influencing policies, ethics, and standards, improving product diversity, consumer education, and public health risk mitigation efforts, and providing the groundwork "for a smarter and more equitable federal legalization when the time comes" [34].

While considering the implications of the Dormant Commerce Clause [34], allowing states the option to participate in interstate commerce activity could preserve preexisting, state-specific programs and priorities focused on the direct success of each state or region's specialized contribution to an overall national market. Oregon and California have already approved policies seeking guidance from the Department of Justice on the agency's position to initiate trade. The advantages of this consolidation would be the ability to leverage economic synergies by releasing excess product to neighboring states, potentially relieving pressures of state-siloed price compression, and allowing individual states to focus on, or defer, specialized production based on geographic advantages or limitations. This proposed system of state vertical disintegration could result in the preservation of profits and players each performing specific tasks to reduce production costs as *Cannabis* prices continue their race to the bottom. As consumer demand increases, "a steady, varied, and world-class supply chain will help thousands of small businesses and potential businesses in retail, distribution, delivery, product development, manufacturing, wellness, etc., in states where it is difficult or inefficient to grow *Cannabis* at scale to get off the ground years sooner, and to differentiate their businesses based on their access to thousands of suppliers" [34]. It might be a good time to be in the "pick and shovel" business.

In the world of golf, equality represents the allowance or invitation to play on the course, while equity represents the utilization of clubs, as well as training and practice. Decades of disproportionate justice and systematic market exclusion cannot be rectified by the mere invitation or allowance to play on the course. The good faith efforts proposed by current social equity programs could do more than merely provide the clubs or tools to play, it could offer minority *Cannabis* business owners the years of training and practice in the form of an "exclusive" head start. As currently proposed by the Craft *Cannabis* Alliance, Congress could allow only those businesses participating in state equity programs to engage in interstate commerce until certain benchmarks measuring fairness are achieved [34]. This exclusive advantage within a "gray" window of opportunity could provide minority *Cannabis* business owners the time and space necessary to have a fair chance of success. Similarly, Shawn Collins, Massachusetts *Cannabis* Control Commission Executive Director, posits a new form of license permitting social consumption in *Cannabis* Cafes available exclusively to social equity applicants for a proposed three-year period [23]. Many combinations of these innovative recommendations could potentially provide those most harmed by the past "shady" political agendas enabling the "war on drugs," an early "tee time," space to develop their "A-game," and an opportunity to "drive the green" prior to federal legalization.

The inconsistent legality of *Cannabis*, as well as unstandardized regulations, among individual states has created an opportunity for illicit market arbitrage placing citizens, both young and old, in danger of potential public health risks. Adulterated *Cannabis* products illicitly manufactured in legal states possessing oversaturated markets could easily be transported physically and sold in states possessing no legal *Cannabis* market regulations or standards at a premium due to product scarcity. The aftermath of this transaction could potentially lead to numerous negative health outcomes including unintentional ingestions by minors or large numbers of illnesses due to unregulated "black" market products. Beyond the hazards of illegally imported edibles or vaping products, the Hemp/CBD market is experiencing its own whirlwind of dangerous confusion. Following the Farm Bill of 2018, presumed legal CBD and hemp products began proliferating gas stations and convenience stores nationwide. As savvy pharmaprenuers experimented with derivative compounds, both natural and synthetic, new innovative "hemp-derived" products began hitting the shelves with the promise of intoxicating effects similar to THC. In response as of February 2023, the Drug Enforcement Agency (DEA) concluded that "delta" cannabinoids, including

Delta-8 THCO and Delta-9 THCO, synthetically derived from hemp do not qualify as hemp, and will be viewed as illegal controlled substances [35]. Concurrently, as the Biden Administration signed the "Medical Marijuana and Cannabidiol Research Expansion Act", the FDA ruled that existing regulatory frameworks for Foods and Supplements will not translate over to Cannabidiol and a new regulatory pathway must be found [36]. While established industries with experience, funding, and preexisting standards await federal legalization, future *Cannabis* industry business practices could emulate these standards to attempt to prevent negative public health outcomes. Until then, regulators will continue to struggle to create standards that make logical sense, while bad actors, endangering public health and providing consumer miseducation, continue to make actual dollars.

Conclusion

At present, numerous nations have legalized the use of *Cannabis* globally, as the U.S. *Cannabis* experiment continues. While prioritizing access to traditional banking and overcoming 280E tax code issues as high-level business practice concerns, inconsistencies among individual state policies, ethics, and standards enabling legal confusion, racially motivated exclusion, and potential public health risks could be addressed with an equal level of urgency. By considering the voluntary incremental allowance of interstate commerce prior to federal legalization, states could choose whether to participate, while retaining their original programs and priorities intact. These "laboratories of democracy" could provide models for future best practices in averting public health risks, as well as disincentivizing production activities in regions deemed noncompetitive following federal legalization. Furthermore, permitting those systematically targeted by the "shady" policies of the past advanced exclusivity within a "gray" window of opportunity could provide individuals a necessary head start to develop a world-class supply chain comprised of retail, distribution, delivery, product development, manufacturing, and wellness, as well as consumption lounges, vacation lodging or other *Cannabis* tourism-based innovations. By inviting and incorporating these previous "black" market participants, millions of consumer dollars could likewise move from the illicit market years prior to federal legalization.

While goldminers, fishermen, snake oil salesmen, and golfers all serve as suitable, relatable allegories, none of these occupations or industries run the risk of being deemed federally illegal. As the U.S. *Cannabis* experiment continues, the potential influence of permitted interstate commerce and a singular uniform concept of acceptable policies, ethics, and standards could lead to the discovery of possible consolidation and streamlining of operations, achievement of stable price equilibrium across all states, and specialized production capitalizing on regional advantages, as well as increased product diversity, consistent consumer education, and the mitigation of associated public health risks.

References

[1] Robertson K. Marijuana industry layoffs and cost-cutting likely to continue in 2023. MJBizDaily, 2023. (Accessed on March 1, 2023, at https://mjbizdaily.com/marijuana-industry-layoffs-and-cost-cutting-likely-to-continue-in-2023/)

[2] Casacchia C, Roberts C, Robertson K Projected US legal medical and recreational cannabis market size, MJBizDaily, 2022. (Accessed March 1, 2023, at https://mjbizdaily.com/us-cannabis-sales-estimates/)

[3] Cannabis Pricing Analysis Shows Commoditization In Fully Legalized States, Weekly stash the cannabis industry. Intro-act, 2022. (Accessed March 1, 2023, at https://www.intro-act.com/Stash/13-04-2022/stash.html#storyoftheweek).

[4] Degenhardt L, Hall W, Lynskey M. The Global Burden of Disease Attributable to Alcohol and Drug Use in 195 Countries and Territories, 1990–2010: A Systematic Analysis. The Lancet 2022;382 (9904):1564–1574.

[5] DeAngelo A The Hockey Stick Turns into Bell Curve: A New Report Sheds Light on Cannabis Industry Growth, Forbes. Forbes Magazine, 2022. (Accessed March 1, 2023, at https://www.forbes.com/sites/andrewdeangelo/2022/10/04/the-hockey-stick-turns-into-bell-curve-a-new-report-from-bdsa-sheds-light-on-cannabis-industry-growth/?sh=37e00f1d66f2).

[6] Kline A, Tobin T. Cannabis and the Dormant Commerce Clause, 2022. (Accessed March 1, 2024, at https://www.perkinscoie.com/en/news-insights/cannabis-and-the-dormant-commerce-clause.html)

[7] Schaneman B. Cannabis Growers in Mature Markets call for License Moratorium Amid Falling Prices. MJBizDaily, 2022. (Accessed March 1, 2024, at https://mjbizdaily.com/cannabis-growers-call-for-license-moratorium-amid-falling-prices/)

[8] Lange T. Year-Over-Year Cannabis Sales Shrink for First Time in 5 States. Cannabis Bus Times, 2022. (Accessed March 1, 2024, at https://www.cannabisbusinesstimes.com/news/yoy-cannabis-sales-shrinking-5-states-2022/)

[9] Marijuana Business Daily. Annual Marijuana Business Factbook: The Definitive Guide to the Cannabis Industry. 7th Edition, 2020. (Accessed March 1, 2023, at https://mjbizdaily.com/bizbooks/factbook-2019/IV/).

[10] Lerman R. How the Marijuana 'Green Rush' Fell Apart. Washington Post, 2022. (Accessed March 1, 2024, at https://www.washingtonpost.com/business/2022/12/30/marijuana-supply-sales-turmoil/)

[11] Cannabis Flower: Examining pricing compression in the US & Canada. Headset, 2022. (Accessed March 1, 2023, at https://www.headset.io/industry-reports/cannabis-flower-examining-pricing-compression-in-the-us-canada)

[12] Pricing Compression in an Evolving Cannabis Market. BDSA, 2022. (Accessed March 1, 2023, at https://bdsa.com/content/pricing-compression-in-an-evolving-cannabis-market-2/)

[13] Maravelias P. Adult-Use Marijuana Companies Struggle to Stay Afloat Amid Overproduction, Falling Prices, MJBizDaily, 2022. (Accessed March 1, 2024, at https://mjbizdaily.com/adult-use-marijuana-companies-struggle-to-stay-afloat/)

[14] Packer C Why We Should Abandon the Term "Black Market", Filter, 2023. (Accessed March 1, 2023, at https://filtermag.org/black-market-cannabis/)

[15] American Civil Liberties Union. The War on Marijuana in Black and White, 2019. (Accessed March 1, 2023, at https://www.aclu.org/report/war-marijuana-black-and-white)
[16] Alexander M. The New Jim Crow: Mass Incarceration in the Age of Colourblindness. Penguin Books, 2019.
[17] Harris KN, Martin W. Persistent Inequities in Cannabis Policy, 2021. (Accessed March 1, 2024 at https://www.bakerinstitute.org/sites/default/files/2021-02/import/inequities-in-cannabis-policy-2021.pdf)
[18] Bailey J. Seeds of Change: Strategies to Create an Equitable Cannabis Industry, Leafly, 2021. (Accessed March 1, 2024 at https://leafly-cms-production.imgix.net/wp-content/uploads/2021/06/25091621/Leafly-2021-white-paper-Seeds-Of-Change-by-Janessa-Bailey-1.pdf)
[19] Title S. Fair and Square: How to Effectively Incorporate Social Equity into Cannabis Laws and Regulations. Buffalo Law Rev 2021;69(4):939–973.
[20] Hinckley JD, Hopfer C. Marijuana Legalization in Colorado: Increasing Potency, Changing Risk Perceptions, and Emerging Public Health Concerns for Youth, Adolescent Psychiatry (Hilversum, Netherlands), 2021. (Accessed March 1, 2024, at https://www.ncbi.nlm.nih.gov/pmc/articles/PMC10078756/)
[21] Disparity in Workforce and Ownership in New York's Medical Cannabis Industry. New York State Office of Cannabis Management, 2023. (Accessed March 1, 2024, at https://cannabis.ny.gov/system/files/documents/2024/01/12-29-23-ocm-report.pdf)
[22] Connley C. Cannabis is Projected to be a $70 Billion Market by 2028 – Yet Those Hurt Most by the War on Drugs Lack Access, CNBC, 2021. (Accessed March 1, 2024, at https://www.cnbc.com/2021/07/01/in-billion-dollar-cannabis-market-racial-inequity-persists-despite-legalization.html)
[23] Christensen D. Social Equity Programs Try to Level Playing Field for Cannabis, Valley Advocate, 2022. (Accessed March 1, 2024, at https://valleyadvocate.com/2022/07/12/social-equity-programs-try-to-level-playing-field-for-cannabis/)
[24] Meadows M. Promoting Safe and Effective Drugs for 100 Years. U.S. Food & Drug Administration, 2006. (Accessed March 1, 2023, at https://www.fda.gov/files/Promoting-Safe-and-Effective-Drugs-for-100-Years-%28download%29.pdf)
[25] Herrington AJ. New Certification Sets Standards for Cannabis Lab Testing, Forbes. Forbes Magazine, 2023. (Accessed April 4, 2023, at https://www.forbes.com/sites/ajherrington/2023/04/04/new-certification-sets-standards-for-cannabis-lab-testing/?sh=2ff111231cb8).
[26] Kim T. Editor: Marijuana Poison Control. LiveStories, 2020, July 21. (Accessed November 2, 2022, at https://insight-editor.livestories.com/s/v2/marijuana-poison-control/e013c91f-c9f1-470c-9e20-03b3b3e9e46c)
[27] Grady D. Dank Vapes, TKO and Other THC Vaping Brands are Linked to Illnesses, C.D.C. Says. Nytimes.com, 2020. (Accessed September 27, 2020, at https://www.nytimes.com/2019/09/27/health/vaping-illness-thc.html.)
[28] Cannabis Laws and Rules. Oregon Liquor and Cannabis Commission: Cannabis Laws and Rules: Marijuana and Hemp (Cannabis): State of Oregon. (Accessed March 1, 2023, at https://www.oregon.gov/olcc/marijuana/pages/recreational-marijuana-laws-and-rules.aspx)
[29] Department of Cannabis Control – State of California. California's cannabis laws, Department of Cannabis Control. (Accessed March 1, 2023, at https://cannabis.ca.gov/cannabis-laws/laws-and-regulations/)
[30] Center USCS Cannabis Regulations Inadequate Given Rising Health Risks of High-Potency Products, USC Schaeffer, 2022. (Accessed April 4, 2023 at https://healthpolicy.usc.edu/article/cannabis-regulations-inadequate-given-rising-health-risks-of-high-potency-products/)

[31] McKeil J The Industry's Dirty Secret: Lab Shopping. Cannabis Technology News, 2022. (Accessed March 1, 2023, at https://www.cannabistech.com/articles/the-industrys-dirty-secret-lab-shopping/)

[32] Gieringer D NORML's Guide to Delta-8 THC and Other Novel Cannabinoids. NORML, 2023. (Accessed March 1, 2023, at https://norml.org/marijuana/fact-sheets/normls-guide-to-delta-8-and-other-novel-cannabinoids/)

[33] Title S Bigger Is Not Better: Preventing Monopolies in the National Cannabis Market. SSRN, 2022. (Accessed March 1, 2024, at https://papers.ssrn.com/sol3/papers.cfm?abstract_id=4018493)

[34] Berke J, et al. Regulating Cannabis Interstate Commerce: Perspectives on How the Federal Government Should Respond, SSRN Electron J [Preprint], 2022. (Accessed March 1, 2023, at https://doi.org/10.2139/ssrn.4188089)

[35] THC-OH NO! The DEA Classifies Delta-8 and Delta-9 THCO as Controlled Substances. Bradley, 2023 (Accessed March 1, 2024, at https://www.bradley.com/insights/publications/2023/02/thcoh-no-the-dea-classifies-delta8-and-delta9-thco-as-controlled-substances).

[36] Woodcock J. FDA concludes that existing regulatory frameworks for foods and supplements are not appropriate for cannabidiol, will work with Congress on a new way forward, U.S. Food and Drug Administration. FDA, 2023. (Accessed March 1, 2023, at https://www.fda.gov/news-events/press-announcements/fda-concludes-existing-regulatory-frameworks-foods-and-supplements-are-not-appropriate-cannabidiol)

Andrew M. Peterson

Health outcomes research in the *Cannabis* industry

Introduction

Outcomes research, as defined by the International Society of Pharmacoeconomics and Outcomes Research (ISPOR), "comprises a set of scientific disciplines that evaluate the effect of healthcare interventions on patients" [1]. Common measures used in health outcomes research include mortality, effectiveness of care, timeliness of care, and others that impact the health status of patients. However, many of these are not primary outcomes in traditional randomized, controlled clinical trials. Instead, these outcomes are often seen as secondary outcomes, or if they are primary, they are seen in observational studies. Thus, the value of these outcomes is often slighted in clinical research and not well recognized as a means of defining true benefit of an intervention. In this same context, *Cannabis* would be considered a "healthcare intervention" but further bears the brunt of criticism since current regulations in the United States and around the world hamper its ability to be truly studied in randomized controlled trials (RCTs), or if it is used in RCTs, used at doses lower than seen in the real world [2].

The purpose of this chapter is to outline how health outcomes research is used to help create the foundation for strengthening the body of evidence for *Cannabis* use.

Overview of health outcomes research

Over the past 30 years, health outcomes research has significantly evolved and has become an integral part of health and medical research. This field focuses on assessing the impact of healthcare interventions and strategies on patient outcomes and quality of life. It has gained prominence due to its ability to provide evidence-based insights that inform clinical decision-making, healthcare policies, and resource allocation.

Health outcomes measures may be clinical measures or patient-reported outcome (PRO) measures. Examples of clinical measures include resolution of disease, changes in objective laboratory values such as cholesterol levels and blood pressures. PROs involve direct input from patients regarding their health status, symp-

Acknowledgments: The author would like to acknowledge the efforts of Arielle Norton, PharmD, Clinical Assistant Professor with Rutgers University and Cooperman Barnabas Medical Center in reviewing this chapter.

Andrew M. Peterson, Philadelphia College of Pharmacy, Saint Joseph's University

https://doi.org/10.1515/9783110677508-008

toms, functional abilities, and quality of life. By incorporating patient perspectives, PROs provide a comprehensive assessment of the effectiveness and impact of medical interventions [3].

Health outcomes research has revolutionized the field of medicine by providing invaluable empirical evidence regarding treatment effectiveness, patient preferences, and healthcare decision-making. While RCTs are widely considered the gold standard in evidence-based research, the unique characteristics of whole plant extract *Cannabis* and the growing acceptance of its medicinal applications in numerous countries have opened doors for the utilization of real-world data (RWD) [4]. In fact, due to the restrictive nature of *Cannabis*, research using RCTs in controlled clinical settings is limited, whereas observational research conducted via dispensaries collects RWD to demonstrate the medical value of the plant.

As many healthcare policies and approvals for *Cannabis* products are based on these outcomes, researchers face an additional responsibility to ensure the quality and rigor of their studies when utilizing real-world evidence (RWE). This demands a meticulous approach to uphold the standards necessary to validate the medicinal use of *Cannabis* and provide substantial evidence for regulatory decision-making.

It is evident that the evolving field of health outcomes research continues to play a pivotal role in shaping healthcare policies and enhancing patient outcomes. RWE offers a unique perspective by examining the effectiveness and impact of treatments in real-world settings, providing valuable insights that complement traditional clinical trials. By exploring the benefits and risks of *Cannabis* as a medicinal option through the lens of real-world data, researchers can contribute to a comprehensive understanding of its therapeutic potential and inform evidence-based healthcare decisions.

Real-world evidence

The US Food and Drug Administration (FDA) defines RWE as "clinical evidence about the usage and potential benefits or risks of a medical product derived from real-world data" [5]. RWD is data derived from electronic health records or claims data that inform about patients' health status. RWE studies are a type of research that examines the safety and effectiveness of medical interventions in a real-world setting, beyond the controlled environment of clinical trials. These observational studies have the advantage of being more generalizable to real-world populations, as they reflect actual clinical practice. However, they are often limited by confounding variables, and the results may be influenced by biases such as selection bias, patient compliance, and incomplete information; thus, causal relationships between an intervention and an outcome may be difficult to decisively conclude [6].

Both RCTs and observational studies are common study designs used in healthcare research. Both are valuable tools to help create the body of evidence needed to demonstrate the effectiveness of a medical intervention, such as a drug. While systematic reviews of clinical trials are at the top of the hierarchy of the evidence pyramid (Figure 1), RCTs are benchmark for experimental research because of the rigor associated with them. The random allocation of treatment, the controlling of confounding variables, and the minimization of threats to study validity are strengths that allow the researchers to conclude, with confidence, the effectiveness or ineffectiveness of the treatment. However, these study strengths contribute to their limitations. Because the experimental conditions of a RCT are so controlled, the generalizability to the world at large is difficult. This is where the observational studies come into play. These studies do not control the experimental conditions, rather they observe how patients behave in the real world. Thus, these studies do not have the same internal validity – that is, because confounding factors and other threats to study validity are not controlled for – the ability to declare the results as being due to the treatment is limited. However, as noted, their strength lies in their ability to capture the real-world actions; thus, the results of these studies are more generalizable.

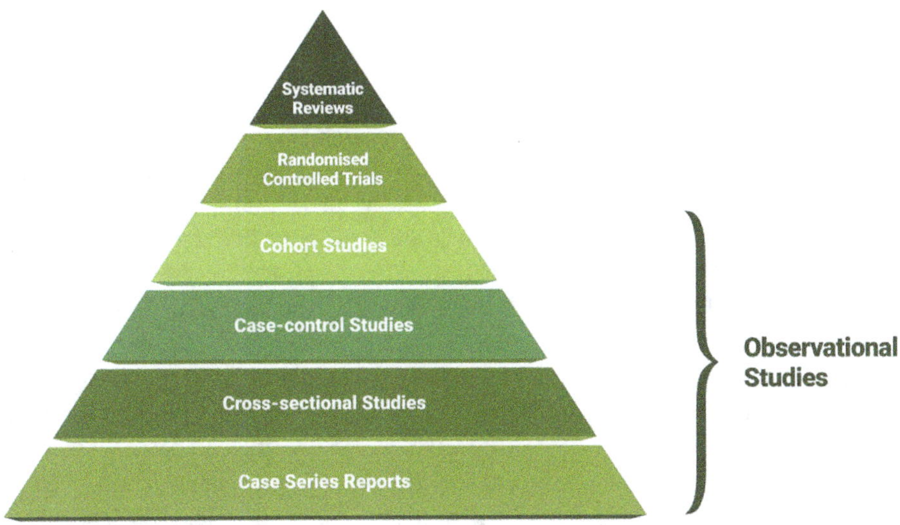

Figure 1: Hierarchy of Study Designs.

RWE studies are becoming increasingly important in healthcare because they provide a more complete understanding of how treatments work in patients. This type of research can reveal how treatments work in diverse populations, including patients who are typically excluded from clinical trials, such as those with multiple comorbidities, those taking multiple medications, and those with varying socioeconomic backgrounds. They are beginning to be used to inform regulatory decisions, guide clinical practice, and identify where areas of research are needed [5, 6]. The potential for RWE in the *Cannabis* space can be seen when examining how patients are faring given specific products or THC concentrations and how they affect their PROs.

Patient-reported outcomes

Data collected directly from patients about their health, symptoms, and quality of life is considered a PRO. These outcomes are becoming more common in today's observational research as RWE evaluating the effectiveness of treatments and interventions. These data give the researcher and clinician an opportunity to see the perspective of the patient and how their disease affects their daily lives. See Table 1 for common domains encompassing PROs.

Table 1: Domains Patient-Reported Outcome Measures.

Functional ability/disability
Health perceptions
Health-related quality of life
Health status
Quality of life
Symptoms

One notable development in health outcomes research is the increased utilization of PROs. In 2009, the FDA released guidance on how PROS can be incorporated into clinical trials as endpoints for drugs seeking approval for marketing within the United States [7]. However, they note that the PROs included must be relevant to the treatment goals and the disease being studied. This document further provides guidance on developing and validating PROs for use in new drug application submissions.

There are several sources of PROs, with the Patient-Reported Outcomes Measurement Information System (PROMIS), developed by the National Institutes of Health [8], being a primary source of standardized, validated questionnaires that measure a variety of domains, including physical function, mental health, and pain. These questionnaires have been used in numerous studies to evaluate the effectiveness of medical interventions. PROs are useful pieces of data that can provide insight into a patient's health status and quality of life. They are increasingly being used in real-world evidence studies and are supported by the FDA for new drug approvals. How they are being used in the *Cannabis* world is the next section of this chapter.

Applications in *Cannabis* research

Using RWE and PROs in *Cannabis* research brings both benefits and challenges. One of the major advantages of RWE and PROs is their ability to capture data directly from patients in real-world settings, providing insights into the actual use and outcomes of *Cannabis* in diverse patient populations. Unlike traditional clinical trials,

RWE *Cannabis* studies are easier to conduct as they do not require the same level of regulatory approval, making them more flexible and adaptable to the rapidly evolving *Cannabis* industry. This allows researchers to gather data on the effects of *Cannabis* as it is currently being used, providing a more realistic representation of its impact in real-life scenarios.

However, one of the challenges associated with using these designs for *Cannabis* research is the potential lack of standardized protocols and quality control measures. Unlike clinical trials, where strict protocols are followed, *Cannabis* studies using RWE rely on data collected from various sources that may vary state to state and even by dispensary within a state. In *Cannabis* research, since much of the data are collected directly from the patients who received their product from a state dispensary, the exact product composition, daily dosing, and even the interpatient variation can introduce bias and uncertainty, thus limiting the ability to draw definitive conclusions. It is important for researchers to carefully consider these factors when analyzing RWE and PROs to ensure the validity and reliability of the findings.

One example of the utilization of RWE and PROs in *Cannabis* research is published by Peterson et al., in which they studied the effect of *Cannabis* on self-reported health-related quality of life (HRQoL) [9]. Over an eight-week period of time, four surveys were administered to individuals self-identifying as patients approved to use *Cannabis* for the treatment of pain in the Commonwealth of Pennsylvania. These patients were recruited through local dispensaries and via mailing lists of users from the Releaf app, a free tool that helps patients track, learn from, and improve their medical *Cannabis* use and experience [10]. Of the 181 study completers, the HRQoL improved by a small, but statistically significant amount and was bolstered by significant improvements in the anxiety and pain subscales. This RWE study found that users had a significant improvement in their HRQoL in addition to improvements in their pain.

Meng et al. conducted a prospective observational study, following participants from community-based *Cannabis* clinics in Ontario, Canada, for up to 12 months [11]. They surveyed the participants on patient-reported outcomes including pain intensity, quality of life, sleep, and anxiety using validated questionnaires, similar to those found in PROMIS. The 12-month findings showed that medical *Cannabis* use is associated with significant improvements in pain intensity (change from 5.85 to 4.76 on a 10-point scale), and the number of individuals self-reporting opioid use declined by more than 50%. The study highlights the potential benefits of medical *Cannabis* as a treatment option for chronic pain management and emphasizes the importance of PROs in assessing the effectiveness of medical interventions.

These examples demonstrate how the integration of RWE and PROs enables researchers to gain insights into the therapeutic effects, safety profiles, and patient experiences associated with *Cannabis* use in various contexts. By leveraging these methodologies, the *Cannabis* industry can continue to advance its understanding of

Cannabis-based treatments and develop evidence-based practices to better serve patients and professionals in the field.

Despite these advantages, there are concerns using PROs in *Cannabis* research. Primarily, pundits typically state that there is "insufficient evidence of efficacy" or "unknown safety" for using *Cannabis* due to the lack of RCTs. This hesitation is due to the limitations of using PROs in research studies. One limitation is the potential for bias or variability in patient responses, as PROs rely on patient self-report and may be influenced by factors such as mood or social desirability [12]. Another limitation is the potential for missing data, as patients may not always complete all PRO measures or may provide incomplete or inconsistent responses [4].

Overall, however, the use of PROs in research studies has the potential to improve medical care by providing a more patient-centered approach to healthcare. PROs can help to identify patient priorities and preferences and can be used to evaluate the effectiveness of treatments and interventions from the patient's perspective. The use of standardized PRO measures, such as PROMIS, can help to overcome some of the limitations of PROs by providing validated and reliable measures that can be used across studies and settings.

Dispensary-based research

When it comes to studying the effects of state-market *Cannabis*, conducting RCTs can be challenging or even impossible. Therefore, it becomes crucial to compare RCTs with alternative study designs. A recent Cochrane review examined 14 systematic reviews, encompassing 1,583 meta-analyses across 228 medical conditions [13]. This comprehensive review concluded that observational studies provide results that are highly consistent with RCTs, albeit with a slight bias toward smaller effects in observational studies. This suggests that well-designed observational trials can be legally employed to study state-market *Cannabis* and yield findings consistent with RCTs [14].

However, it's important to acknowledge that observational trials often face limitations, particularly in measuring the degree of exposure to the drug. Since observational studies, particularly in *Cannabis* research, rely on self-reported usage, the amount of active ingredient actually consumed is difficult to determine. Furthermore, the significant variation in product quality and active ingredient content varies dramatically, adding to the complexity in determining the level of exposure to specific cannabinoids. This aspect is rarely addressed in the study design. Therefore, the inability to accurately measure and account for these factors can weaken the results of observational trials [14].

Medical *Cannabis* research frequently involves dispensaries, state-licensed establishments that provide *Cannabis* for medicinal use. These dispensaries play a crucial role in research studies as they have access to patients who use *Cannabis* for medical purposes, allowing for data collection in real-world settings. In recent years, numerous studies conducted by dispensaries have been published, shedding light on the effectiveness and benefits of medical *Cannabis*.

One notable example is a study published which examined the use of *Cannabis* for chronic pain management [15]. Accessing patients via websites, flyers in dispensaries, or word of mouth from primarily the US (77.8%), UK (5.6%), and Canada (3.1%), the researchers collected a sample of more than 2,400 eligible respondents. While they admit that generalization about the results should be used with caution due to the convenience sample and self-selection, the results show a willingness of *Cannabis* users to participate in research on the product.

While conducting research through dispensaries offers advantages, there are potential drawbacks to consider. One concern revolves around the financial incentives dispensaries may have to promote the use of *Cannabis* for medical purposes, which could introduce bias into research findings. Additionally, dispensaries may lack the same level of scientific expertise found in academic research settings, potentially compromising the rigor and standards of their studies. Another issue lies in the representativeness of the patient population participating in dispensary-based research. Dispensaries may attract a specific subset of patients, potentially limiting the generalizability of findings to a broader population of individuals using *Cannabis* for medical reasons. Future studies will have to start differentiating between recreational and medical users and explore how that affects RWE outcomes.

In conclusion, dispensaries serve as valuable partners in medical *Cannabis* research, leveraging their access to patients and RWD. However, it is essential to address potential biases and limitations associated with financial incentives and patient selection to ensure the reliability and generalizability of the research findings. Moving forward, collaboration between dispensaries and academic researchers can help navigate these challenges, ultimately advancing our understanding of medical *Cannabis* and its applications.

Cannabis research opportunities and challenges

Cannabis has been used as a medicine for thousands of years and has been the subject of increasing research interest in recent decades. However, *Cannabis* is classified as a Schedule I drug under the US federal government's Controlled Substances Act, which makes it difficult to conduct research on its potential medicinal benefits. Researchers who want to study *Cannabis* in a clinical trial must obtain approval from several federal agencies, including the Drug Enforcement Administration (DEA), the National Institute on Drug Abuse (NIDA), and the Food and Drug Administration (FDA). This multistep process requires researchers to comply with federal research protocols, security measures to prevent diversion and use the federal government's source of *Cannabis* for the research.

The FDA plays a crucial role in authorizing clinical trials, requiring sponsors to submit an Investigative New Drug (IND) application outlining study protocols, certificates of analysis for cannabinoid profile and purity, stability testing, manufacturing procedures, and investigator qualifications. Researchers must also ensure informed consent, subject protection, and secure storage of *Cannabis*, since it is a controlled substance. The DEA registration is another requirement, with researchers applying under Section 823(f) of the Controlled Substances Act (CSA), providing information about the IND, investigators, research institution, institutional review board (IRB) approval, state licensure, and security protocols. Researchers have faced administrative challenges and delays in obtaining DEA registrations.

However, recently, the DEA announced that it would be taking steps to improve access to marijuana research [16]. The changes promulgated by the DEA in 2020 now allow for additional entities to register as "bulk manufacturers" to cultivate and supply marijuana for research. This announcement was seen as a significant step forward for researchers who have long struggled to obtain high-quality *Cannabis* for their studies due to the limited supply from the federal government. This increase in the number of registered producers is expected to increase the availability of high-quality marijuana for research. However, researchers still need to navigate the challenges of obtaining a Schedule I registration, including administrative hurdles and potential discretionary denial by the DEA based on public health and safety considerations. This process can take years, and some researchers have reported that the regulatory requirements have prevented them from conducting studies altogether.

The DEA's announcement was welcomed by researchers and advocates who have long called for improved access to marijuana for research purposes. While this does not change marijuana's status as a Schedule I substance, this will help advance research on *Cannabis* and expand sources of research-grade marijuana, leading to increased commercial development of FDA-approved *Cannabis* products and help doctors discuss the potential risks and benefits of marijuana and CBD with their patients. Overall, the announcement was seen as a positive development for marijuana research, but there is still much work to be done to improve access to this important area of study.

The classification of *Cannabis* as a Schedule I drug poses significant barriers to conducting research using state-regulated *Cannabis* products. One challenge arises from legislation that grants the federal government the authority to impose punitive measures, such as the withdrawal of all federal funding, against universities involved in research with products that are legally accessible within their respective states. As a result, even in states where *Cannabis* has been legalized, scientists affiliated with universities are unable to investigate the potential health risks and benefits of these products through conventional research methods. This restriction hinders the exploration of crucial public health implications that could impact residents within the state [14].

In April 2022, the House of Representatives passed the Marijuana Opportunity Reinvestment and Expungement (MORE) Act, which would remove *Cannabis* from the list of Schedule I drugs and eliminate many of the federal barriers to *Cannabis* research [17]. This would, in turn, allow for companies to conduct research on commercially available *Cannabis* products using traditional RCTs, thus expanding the internal validity and potentially the generalizability while improving the quality of data on the value of *Cannabis* as a medicinal product. Despite these efforts, the regulatory environment still poses significant challenges to *Cannabis* research. For example, if *Cannabis* was rescheduled to a Schedule II, businesses would still be subject to the 280E tax regulations, disallowing certain deductions and limiting the profitability of the organization [18].

Some critics argue that the federal government's monopoly on the production of marijuana for research purposes creates a conflict of interest [19]. The federal government has a vested interest in maintaining the Schedule I classification of marijuana, which makes it difficult for researchers to conduct studies on its potential therapeutic benefits. Critics argue that this conflict of interest could lead to bias in the research that is conducted and could prevent important discoveries from being made. For example, the FDA has not yet approved any *Cannabis* products for medicinal use [5] and there is still a lack of high-quality clinical research on the safety and efficacy of *Cannabis* for various medical conditions. Note that the FDA has approved one *Cannabis*-derived product and three synthetically manufactured THC products (see Table 2), but these products do not have the clinical impact nor the appeal that plant-based products have to the general public. Further, the FDA-approved products are not indicated for the treatment of many of the conditions for which medical *Cannabis* is used (e.g., pain) and, therefore, leaves the medical world little choice but to use plant-based products.

Table 2: FDA-Approved *Cannabis*-Related Products.

Cannabidiol (Epidiolex®) – plant derived
Dronabinol (Marinol®) – synthetic
Dronabinol (Syndros®) – synthetic
Nabilone (Cesamet®) – synthetic

One significant hurdle for researchers is the quality of the *Cannabis* that is available for research. Until recently, the University of Mississippi was the only federally approved source of legal *Cannabis* for research in the United States, and the *Cannabis* it provided was criticized for poor quality and not an accurate reflection of the potency and variety of the *Cannabis* products available in the marketplace [2, 20, 21]. This makes it difficult for researchers to study the effects of different strains of *Cannabis* and to determine the optimal dosages for various medical conditions, severely limiting the ability of federally obtained *Cannabis* to be useful in real-world settings [22].

In 2017, NIDA commissioned a workgroup to develop a research agenda for *Cannabis*. In their 2018 report, they noted that measuring standardized doses of *Cannabis* across its various product forms and modes of administration is crucial for understanding exposure, intoxication, and impairment. Because past research predominantly used low-THC *Cannabis* and current market products include higher-potency items and newer methods of administration, there is the potential for delivering higher doses of THC. However, the lack of a comparable concept or measure for *Cannabis* due to its diverse products, formulations, and forms poses a growing problem. As such, they, and others, have been pushing for a standardized reference dose [23].

Conclusion

In conclusion, while there is growing interest in the potential medicinal benefits of *Cannabis*, the federal regulatory environment and the quality of the *Cannabis* available for research continue to hinder progress in this field. Further research is needed to better understand the therapeutic effects of *Cannabis* and to develop safe and effective *Cannabis*-based treatments. Meanwhile, observational research, using RWE and PROs particularly in dispensary-based research, is a growing mechanism for understanding the beneficial and hazardous effects of *Cannabis*.

References

[1] International Society of Pharmacoeconomics and Outcomes Research. About HEOR, 2023. Retrieved from https://www.ispor.org/heor-resources/about-heor#:~:text=Outcomes%20research%20comprises%20a%20set,of%20healthcare%20interventions%20on%20patients. Viewed 4/11/2023.

[2] National Academies of Sciences, Engineering, and Medicine (NASEM); Health and Medicine Division; Board on Population Health and Public Health Practice; Committee on the Health Effects of Marijuana: An Evidence Review and Research Agenda. 15, Challenges and Barriers in Conducting Cannabis Research, The Health Effects of Cannabis and Cannabinoids: The Current State of Evidence and Recommendations for Research. Washington, DC: National Academies Press (US), 2017 Jan 12. Available from https://www.ncbi.nlm.nih.gov/books/NBK425757/.

[3] Weldring T, Smith SM. Patient-Reported Outcomes (PROs) and Patient-Reported Outcome Measures (PROMs). Health Serv Insights 2013 Aug 4;6:61–68. doi: 10.4137/HSI.S11093. PMID: 25114561; PMCID: PMC4089835.

[4] Sakal C, Lynskey M, Schlag AK, Nutt DJ. Developing a Real-world Evidence Base for Prescribed Cannabis in the United Kingdom: Preliminary Findings from Project Twenty21. Psychopharmacology (Berl) 2022 May;239(5):1147–1155. doi: 10.1007/s00213-021-05855-2. Epub 2021 May 10. PMID: 33970291.

[5] U.S. Food & Drug Administration. Real-World Evidence, 2023. Retrieved from https://www.fda.gov/science-research/science-and-research-special-topics/real-world-evidence. Viewed 4/11/2023.

[6] Kim HS, Kim JH. Proceed with Caution When Using Real World Data and Real World Evidence. J Korean Med Sci 2019 Jan 16;34(4):e28. doi: 10.3346/jkms.2019.34.e28. PMID: 30686950; PMCID: PMC6345637.

[7] U.S. Food & Drug Administration. Patient-Reported Outcome Measures: Use in Medical Product Development to Support Labeling Claims, 2009. Retrieved from https://www.fda.gov/regulatory-information/search-fda-guidance-documents/patient-reported-outcome-measures-use-medical-product-development-support-labeling-claims. Viewed 4/11/2023.

[8] Cella D, Riley W, Stone A, Rothrock N, Reeve B, Yount S, Amtmann D, Bode R, Buysse D, Choi S, Cook K, Devellis R, DeWalt D, Fries JF, Gershon R, Hahn EA, Lai JS, Pilkonis P, Revicki D, Rose M, Weinfurt K, Hays R, PROMIS Cooperative Group. The Patient-Reported Outcomes Measurement Information System (PROMIS) Developed and Tested its First Wave of Adult Self-reported Health Outcome Item Banks: 2005–2008. J Clin Epidemiol 2010 Nov;63(11):1179–1194. doi: 10.1016/j.jclinepi.2010.04.011. Epub 2010 Aug 4. PMID: 20685078; PMCID: PMC2965562.

[9] Peterson AM, Le C, Dautrich T. Measuring the Change in Health-Related Quality of Life in Patients Using Marijuana for Pain Relief. Med Cannabis Cannabinoids 2021 Aug 12;4(2):114–120. doi: 10.1159/000517857. PMID: 35224431; PMCID: PMC8832252.

[10] RELEAF. https://releafapp.com/about-us/. Viewed 6/13/2023.

[11] Meng H, Page MG, Ajrawat P, Deshpande A, Samman B, Dominicis M, Ladha KS, Fiorellino J, Huang A, Kotteeswaran Y, McClaren-Blades A, Kotra LP, Clarke H. Patient-reported Outcomes in Those Consuming Medical Cannabis: A Prospective Longitudinal Observational Study in Chronic Pain Patients. Can J Anaesth 2021 May;68(5):633–644. English. doi: 10.1007/s12630-020-01903-1. Epub 2021 Jan 20. PMID: 33469735.

[12] Chang EM, Gillespie EF, Shaverdian N. Truthfulness in Patient-reported Outcomes: Factors Affecting Patients' Responses and Impact on Data Quality. Patient Relat Outcome Meas 2019 Jun 11;10:171–186. doi: 10.2147/PROM.S178344. PMID: 31354371; PMCID: PMC6573779.

[13] Anglemyer A, Horvath HT, Bero L. Healthcare Outcomes Assessed with Observational Study Designs Compared with those Assessed in Randomized Trials. Cochrane Database Syst Rev 2014 Apr 29; 2014(4):MR000034. doi: 10.1002/14651858.MR000034.pub2. PMID: 24782322; PMCID: PMC8191367.

[14] Hutchison KE, Bidwell LC, Ellingson JM, Bryan AD. Cannabis and Health Research: Rapid Progress Requires Innovative Research Designs. Value Health 2019 Nov;22(11):1289–1294. doi: 10.1016/j.jval.2019.05.005. Epub 2019 Oct 18. PMID: 31708066.

[15] Sexton M, Cuttler C, Finnell JS, Mischley LK. A Cross-Sectional Survey of Medical Cannabis Users: Patterns of Use and Perceived Efficacy. Cannabis Cannabinoid Res 2016 Jun 1;1(1):131–138. doi: 10.1089/can.2016.0007. PMID: 28861489; PMCID: PMC5549439.

[16] U.S. Drug Enforcement Administration. DEA proposes process to expand marijuana research in the United States, 2020. Retrieved from https://www.dea.gov/press-releases/2020/03/20/dea-proposes-process-expand-marijuana-research-united-states. Viewed 5/17/2023

[17] Muriel M. Cannabis legislation update: the MORE Act passes in the House. Retrieved from https://www.bakertilly.com/insights/Cannabis-legislation-update-the-more-act-passes-in. Viewed 5/7/2023.

[18] Freeman JB. Section 280E and The Taxation of Cannabis Businesses, 2022. https://freemanlaw.com/section-280e-and-the-taxation-of-*Cannabis*-businesses/. Viewed 6/8/2023

[19] Kaufmann M. Federal Marijuana Monopoly Challenged, Washington Post, 2005 Dec 12. https://www.washingtonpost.com/archive/politics/2005/12/12/federal-marijuana-monopoly-challenged/5f2b8928-f093-47bc-aced-a2263a68f235/. Viewed 6/8/2023.

[20] Cooper ZD, Abrams DI, Gust S, Salicrup A, Throckmorton DC. Challenges for Clinical Cannabis and Cannabinoid Research in the United States. J Natl Cancer Inst Monogr 2021 Nov 28;2021(58):114–122. doi: 10.1093/jncimonographs/lgab009. PMID: 34850896; PMCID: PMC8783595.

[21] National Institute for Drug Abuse (NIDA): National Advisory Council on Drug Abuse Workgroup. Recommendations for NIDA's Cannabis Policy Research Agenda. National Institute for Drug Abuse (NIDA), 2018.

[22] Vergara D, Bidwell LC, Gaudino R, Torres A, Du G, Ruthenburg TC, deCesare K, Land DP, Hutchison KE, Kane NC. Compromised External Validity: Federally Produced Cannabis Does Not Reflect Legal Markets. Sci Rep 2017 Apr 19;7:46528. doi: 10.1038/srep46528. PMID: 28422145; PMCID: PMC5395929.

[23] Freeman TP, Lorenzetti V. A Standard THC Unit for Reporting of Health Research on Cannabis and Cannabinoids. Lancet Psychiatry 2021 Nov;8(11):944–946. doi: 10.1016/S2215-0366(21)00355-2. Epub 2021 Sep 7. PMID: 34506750.

Teresa A. Simon

How to keep a product on the shelf: a public health approach

Introduction

First, do no harm (or "primum non nocere"). This is the oath that medical students must take before becoming a doctor. Also known as the Hippocratic Oath, it is an ethical standard that dates back to 275 AD. For doctors, abiding by this standard means that they are recommending an action or a treatment that they think is helpful or useful. The original oath contains the phrase "I will abstain from whatever is deleterious or mischievous." Physicians are held to this standard and the doctor–patient relationship is built on trust – trust that the physician will keep the interest of the patient of utmost importance.

This idea of "do no harm," however, expands far beyond medical professionals who explicitly take the oath. Consider the Federal Aviation Agency (FAA) mission statement "to provide the safest, most efficient aerospace system in the world." The Food and Drug Administration (FDA) mission states that it "is responsible for protecting the public health by ensuring the safety, efficacy, and security of human and veterinary drugs, biological products, and medical devices; and by ensuring the safety of our nation's food supply, cosmetics, and products that emit radiation." Occupational Safety and Health Administration (OSHA) mission is "to ensure safe and healthful working conditions for workers by setting and enforcing standards and by providing training, outreach, education and assistance." The Consumer Product Safety Commission (CPSC) formed in 1972 with a mission to "protect the public against unreasonable risks of injury or death from consumer products through education, safety standards activities, regulation, and enforcement."

The concept of safety and protection is woven into the very fabric of the United States regulatory agencies and committees. Does one interpret "do no harm" as protecting and keeping safe the citizens of the United States? What is the difference between feeling safe and being safe? The past two decades have seen an explosion of over-the-counter products and the last 2–3 years have witnessed an implosion of hemp-derived products including cannabidiol (CBD) as well as other products derived from CBD. The public is getting a chemistry lesson; Delta-8 THC, THC-O, THC-V. Okay, hold on. Let's back up. First, we must focus on understanding some basics of public health and safety as well as the history of *Cannabis* and hemp to truly explore the magnitude of this potential public health issue.

Teresa A. Simon, Physicians Research Center, LLC

https://doi.org/10.1515/9783110677508-009

A brief history of public health overall and in America

Public health is a search for effective means of securing health and preventing disease in a population. Prevention of disease is what we are most familiar with. Epidemic and endemic infections have always been at the root of public health, and we are continually faced with similar challenges, most recently the COVID-19 pandemic. Prevention of disease in populations seems relatively straightforward: define the disease, measure how often it occurs, and in whom. After we assess these variables, we seek effective interventions.

Public health evolves through trial and error and is influenced by war, natural disasters, religion, and social systems. A growing need for organized health protection occurs as populations grow and urbanization occurs. Certain groups have influenced approaches to education and have managed to control communicable diseases through sanitation as well as increasing preventative health care. Education is viewed by some as threatening, resulting in the inability to implement change. However, shouldn't everyone be entitled to have basic knowledge of available products to allow consumers to make their own informed decision? Education and safety information is available on the usage of seat belts, smoking of tobacco, as well as vaccinations. Historical public health examples of controlling disease and sanitation include John Snow and the cholera outbreak that devastated London in the mid-nineteenth century, summarized in *The Ghost Map: The Story of London's Most Terrifying Epidemic – And How It Changed Science, Cities, and the Modern World* by Steven Johnson [1]. The meat-packing industry within the United States also went through a period of public health upheaval, as depicted in Upton Sinclair's *The Jungle* [2]. Current examples of organized health protection include immunization, responses to the opioid crisis, and prevention of sexually transmitted diseases such as HIV. However, the list continues to grow, as modern-day public health issues continue to impact societies, regardless of development. Communicable diseases still occur while modern pandemics include cardiovascular disease, cancers, mental illness, and trauma. Furthermore, dangers such as the opioid crisis, maintaining a safe food supply as well as medicines continually challenge us to seek new ways of educating, monitoring products, and preventing the spread of disease and illness. These continued threats have the potential for devastating consequences to society.

The US government's approach to public health has consistently been fractioned. As a result, medical care, public health, and mental health services have evolved separately in terms of policies and procedures. It was not until the mid-to-late 1800s during which the Agricultural Division in the Patent Office led to a federal consumer protection agency with the 1906 Pure Food and Drugs Act. The Pure Food and Drug Act of 1906 prohibited the manufacturing, sale, or transporting of any misbranded, adulterated, poisonous, or deleterious food, drugs, medicines, or liqueurs for other purposes than the primary reason specified on the label [3]. This piece of legislation paved the way for the signing of the Federal Food, Drug and Cosmetic Act in 1938.

There are two additional types of products worth mentioning: botanical drug (natural) products and dietary supplements. A botanical drug product can be made of plant material, algae, or macroscopic fungi. It is a product that can be available as a tea, powder, capsule, topical or injection, and should have an intended use for diagnosis, cure, treatment, or prevention of disease in humans. A few examples of products made from botanicals include aspirin made from Willow bark dating back to use by the Sumerians and Egyptians, digoxin made from foxglove (*digitalis lanata*) to treat conditions of the heart, psyllium a laxative, and St John's Wort to treat mild depression to name a few of the 250,000 flowering plants of which approximately 85,000 have been used medicinally [4].

There are numerous botanical preparations that are marketed in the United States as dietary supplements. A dietary supplement is intended to supplement the diet and is ingested orally (by mouth). The ingredients can include vitamins, minerals, herbs, or other botanicals. The Dietary Supplement Health and Education Act (DSHEA) of 1994 puts dietary supplements in a category under the general umbrella of foods, not drugs. Every supplement must be labeled a dietary supplement and describe how the dietary ingredient affects humans. The labeling on a dietary supplement cannot make a disease claim to diagnose, cure, mitigate, treat, or prevent disease [5].

Pharmaceuticals, botanicals, food products, and additives, e.g., high-intensity sweeteners, alcohol, and tobacco, are all regulated and have their respective guidance. There are numerous public health examples that highlight the importance of regulatory bodies and their processes. The following section outlines two public health examples of products removed from the market: thalidomide, a synthetically derived product, and ephedrine, synthetically derived from ephedra a plant.

Thalidomide is an example of a public health scandal. Thalidomide is a synthetic derivative of glutamic acid. Glutamic acid is one of the 20 natural amino acids that is present in almost every living plant cell. It is the most abundant amino acid in nature and is present in most of the food we eat. Most consumers are unaware that glutamic acid occurs naturally; however, in the late 1950s/early 1960s, the use of thalidomide was worldwide. Thalidomide was marketed as a tranquilizer as well as a medicine to treat anxiety, insomnia, and morning sickness [6, 7]. It was introduced as a sedative and medication for morning sickness without having been tested on pregnant women [8]. While initially deemed to be safe in pregnancy, concerns regarding birth defects were noted in 1961, and the medication was removed from the market in Europe that year [6, 9]. Women exposed to thalidomide during pregnancy had an increased chance of having a child with a birth defect. The thalidomide incident is known as the "biggest man-made medical disaster ever," with more than 10,000 children born with a range of severe deformities, such as phocomelia, as well as thousands of miscarriages [10, 11].

Ephedra/ephedracea (aka Bluestem joint fir) is one of the oldest plants and has been available for centuries in places like China and India to treat colds, fever, headaches, coughing, and wheezing. It is also known as ma-huang and its use dates back to more than 5,000 years ago. Ephedra was used to synthetically create ephedrine. In the United States, ephedra/ephedrine was promoted to increase energy and enhance athletic performance as well as weight loss. Ephedrine was removed from the US market in 2004 after 18,000 adverse events were reported. As Palmer summarizes,

> Ephedrine managed to pass through numerous regulatory loopholes within seventy years. Despite warnings of misuse over the latter half of the century, ephedrine, and its herbal source, ephedra, were regulated in a piecemeal fashion and remained easily available to the public. Health authorities have struggled to control ephedrine, as an amphetamine "look-alike," as a methamphetamine precursor, as a dietary supplement, and as a medication. Despite being a potentially dangerous stimulant, under-regulation was perhaps more problematic than the substance itself.

He concludes that "Tighter control of all ephedrine products, drugs, and dietary supplements alike, might have prevented adverse outcomes and allowed this substance to remain available in a safer manner. Stringent regulation of all ephedrine products is necessary to prevent misuse and to protect the public's health" [12]. Zell-Kanter reports in a letter to the New England Journal of Medicine editor that the 2004 FDA ban proved to be effective in limiting the availability of ephedra and therefore reduced the number of toxicities within the United States [13]. So what is the *Cannabis*/hemp story?

Brief history of *Cannabis*: globally and in the United States

Cannabis sativa L. (C. sativa) is an annual dioecious plant, which shares its origins with the inception of the first agricultural human societies in Asia [14]. Its use by ancient civilizations as a source of food and textile fibers dates back over 10,000 years [15]. Indian use of *Cannabis* dates back to Vedic times, between 1500 BC and 600 BC, and its use was mostly for ritualistic and religious purposes, as documented in the ancient literature [16]. Some societies have used it for rituals, while others for certain ailments. The reasons for use were discovered by trial and error. *Cannabis* and hemp have been in the United States for over 500 years. Some say it was brought over during the Mexican revolution, while others say hemp was brought into North America by Spanish missionaries and colonists. Columbus' landing party documented that they planted it in Chile in the 1490s [17].

There was a time (1850s) when *Cannabis* was available in pharmacies after William O'Shaughnessy introduced the medicine around 1839 [18]. There was even a time when physicians could "prescribe" *Cannabis* as it was listed in the US Pharmacopeia. Today only in states where medical *Cannabis* is legalized can a physician apply to the Department of Health to "recommend" a *Cannabis* product to a patient for an approved condition.

Hemp has had its own commercialization journey, having been grown by landowners initially to support England and later the American colonists grew it to support their own expansion [19, 20]. George Washington and Thomas Jefferson grew hemp as one of their crops [21]. *Cannabis*/hemp cultivation played a central role in the establishment of the United States. The use of hemp for rope and fabric later became ubiquitous in the eighteenth and nineteenth centuries.

Cannabis/marijuana/hemp probably have had the longest history of being caught in a legal and political tug of war. This tug of war continues after decades of the push and pull since 1937 where *Cannabis* continues to be in a legal and regulatory conundrum. With the rhetoric of caring for safety and the public's health, one wonders what the driving force of this industry is.

In 2018, hemp was decoupled and legalized from *Cannabis* "marijuana." In the United States, hemp is defined as a *Cannabis* plant that produces and contains less than 0.3% THC. The statute remained silent on isomers (Delta-8 THC) or synthetically

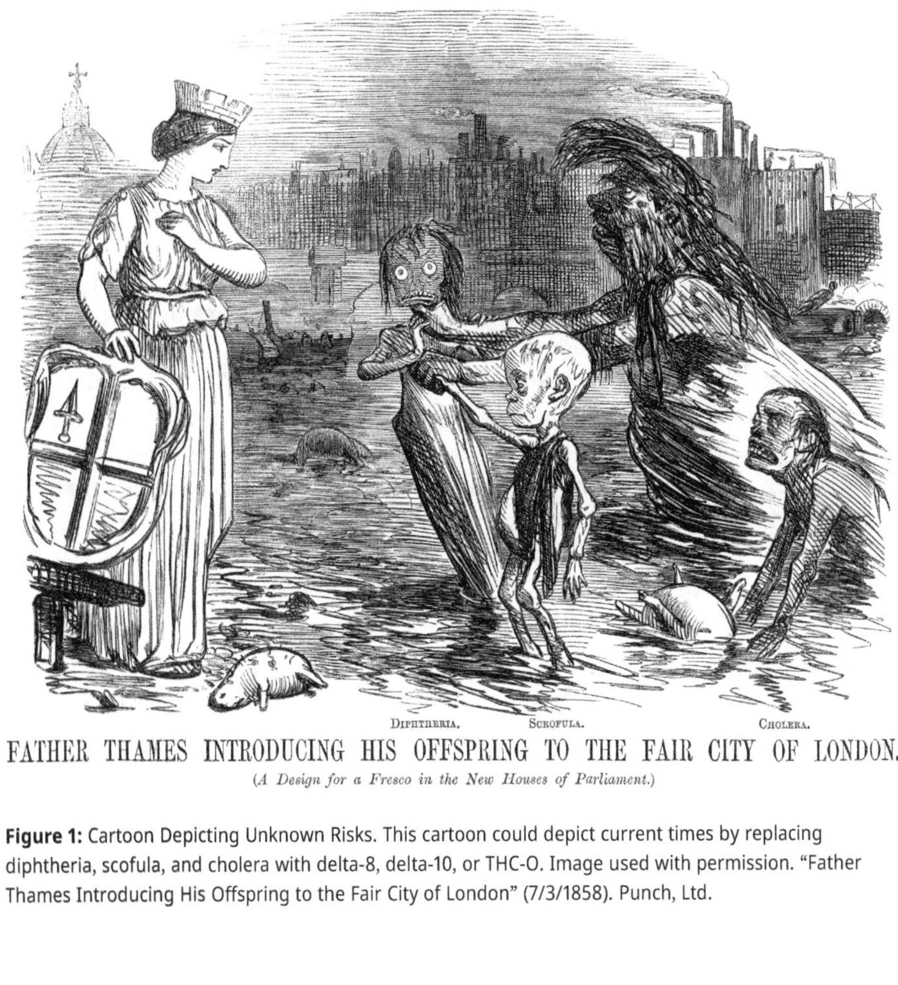

Figure 1: Cartoon Depicting Unknown Risks. This cartoon could depict current times by replacing diphtheria, scofula, and cholera with delta-8, delta-10, or THC-O. Image used with permission. "Father Thames Introducing His Offspring to the Fair City of London" (7/3/1858). Punch, Ltd.

derived (THC-O, THC-V) *Cannabis* products. The legislation did not explicitly state that certain isomers of THC are still Schedule I or which THC-isomer products exceed the 0.3% THC limit. Under this "0.3.% rule," hemp edibles can exceed the THC limit imposed for *Cannabis* edibles (e.g., 10 mg THC maximum in most jurisdictions for *Cannabis* products), as hemp is determined by the percent weight of THC – not a specific amount per dose or unit. Basically, you may be able to buy more edible THC at a gas station than you are able to at a licensed *Cannabis* dispensary (Are you feeling safe now?). "Natural" products such as hemp, *Cannabis*, and ephedra can result in abuse as well as poisonings. There are many cannabinoids that occur naturally and in small amounts on a *Cannabis*/hemp plant including Delta-8 THC. However, a chemical treatment, synthesis, or a change in the chemical structure alters the product and may result in safety issues. Are we experiencing another ephedra to ephedrine situation with the synthetic-hemp-derived products (Delta-8, Delta-10, THC-V, and THC-O)? (Figure 1)

Cannabis and public health

Almost immediately after the legalization of hemp because of the 2018 Hemp Farm Bill, chemists began cooking up hemp-derived products such as Delta-8 THC, THC-O, THC-V, and Delta-10, which seem to be more harmful than the *Cannabis* plant in its raw form [22]. An example of where the raw form of a plant may be safer and provide health benefits compared to its illegally produced synthetic version. When a synthetic formulation is created, there is typically limited or no safety and efficacy testing; therefore, consumers are at risk, leading to yet another public health risk that could be avoided. (This excludes products that have undergone the FDA approval process (e.g., dronabinol).) So, in the middle of the COVID pandemic (2019–2021) when new COVID variants were announced every other day, a new hemp-derived isomer, or minor synthetic chemical variant (THC-O, THC-V, Delta-10 THC), was released to the consumer. As with any new product available to consumers, it may take a trial-and-error period for the benefits and risks to arise. In 2022, a student commented to their fellow classmates in a *Cannabis* education program that I was conducting, "Do you hear this people? We are the guinea pigs." It is sad and true. There are products constantly skirting the processes in place and the public is the last to know. One such example is Delta-8 THC.

Cases of Delta-8 THC poisonings were presented in the emergency rooms and poison control centers across the nation. A recent publication highlighted the toxicities or adverse effects reported in persons using Delta-8 THC products. The authors used one of the surveillance reporting systems put in place (FAERS) to monitor a product's safety [23]. There were enough cases that both the CDC and FDA issued health alerts [24, 25]. Yet, these warnings did not reach mainstream media, as evidenced by another study using social media purported "free weed that gets you high" [26].

Currently, the public health approach to the use of *Cannabis* is framed in the context of mental health, abuse, and addiction. The mild hallucinogenic effects of *Cannabis* have made it a relatively popular drug in the United States as a means of self-medication. The 2020 Substance Use and Mental Health Indicators in the United States: Results from the 2020 National Survey on Drug Use and Health (SAMSA) reported over 50 million persons used *Cannabis* in the past year [27]. Current regulatory authorities have it classified as a drug like heroin and LSD. There are several concerns from a public health perspective, a clear concern is the potential harm unregulated *Cannabis* and hemp products create for the consumer. The hemp industry overproduced CBD products, and the chemists came to the rescue creating derived compounds from a "legal" product creating a public health threat. The public thinks it has access to a safe product – advertised as "free weed that gets you high," but the story continues down a concerning road. Presently, few states have issued regulations banning the selling of Delta-8 THC hemp products. Is there a way out? In my opinion, yes. Hemp should be manufactured as originally intended. Hemp should remain a farming product and produce rope, fabric (textiles and clothing), paper, food, bioplastics, and biofuel. The Department of Agriculture is not in the medicine-producing market. Its mission is to provide leadership on food, agriculture, and natural resources. At the time of this writing, regulated *Cannabis* products are available in 38 states. Briefly, regulated means the companies that sell regulated products must follow good manufacturing processes, which include testing, documenting a chain of custody, and having a label listing the contents of the product. Legalizing *Cannabis* at the national level with dosing limits and clear package labeling including reasons for use, similar to botanicals, could put the hemp industry out of business or force them to be compliant. Hemp is for farming and *Cannabis* is for medicinal use. The *Cannabis* industry can make legitimate products including CBD.

Feeling safe and being safe are not the same

To be safe, we need to be free from harm. There are many aspects that may influence our sense of safety. Some of these aspects are covered in *Perceived Safety* (2019), in which authors Eller and Frey discuss how groups can heavily influence perceived safety [28]. An example includes how the use of helmets in skiers was adopted: initially if a helmet was worn you were not part of the group norm, but then this norm shifted so that wearing a helmet became the new normal. The perception of risk changed even if the actual risk did not change. Our experiences guide us as well as our perception of organizations or situations that ensure our safety.

With so many products on the shelf in the marketplace, it is hard to know what is safe. We assume that if a product is for sale, it must be safe, and that someone is overseeing the safety of the products and therefore the public. As outlined earlier, to some extent this is true, however, not entirely. To be clear, compound medications are products that combine, mix, or alter the ingredients of one or more drugs to create another drug or product. These medications may contain substances that have not been tested for safety or effectiveness and have not been reviewed or approved by the FDA.

As we read earlier, food, botanicals, and dietary supplements can be marketed without obtaining prior approval from the FDA. Products must be registered with the FDA if the product is sold in the United States. Registration requires labeling, a facility, and documentation of good manufacturing facilities. Hemp and CBD products have fallen into this gray zone. Companies are capitalizing on the loophole and the public's sense of feeling safe. To reemphasize, we are not speaking about the regulated *Cannabis* market. Medical *Cannabis* is highly regulated and has testing and labeling requirements. Hemp and hemp-derived products to date do not have to meet the same requirements. This may change before this piece gets to print as this market changes rapidly.

Monitoring

For products not "approved" by the FDA, there are several avenues for consumers to report an adverse outcome or unwanted effect associated with using any product purchased. Your contribution helps the public move closer to *being* safe and forces companies to produce quality products. Safety issues are a process of identifying the abnormal in a sea of normal. Poison control centers, MedWatch, the safety reporting portal, the CDC, and state health departments all have a role in monitoring and reporting any untoward event that may result in a public health threat. Formal surveillance systems such as the Drug Abuse Warning Network (DAWN) was created (1974) to generate real-time information regarding new patterns of drug abuse or the possible importation of contaminated products. The FDA CFSAN Adverse Event Reporting System (CAERS) is a postmarketing surveillance system that receives and monitors adverse event and product complaint reports for foods, dietary supplements, cosmetics, and infant formula. New programs like PulseNet, along with the FDA and USDA, report recalls on food products and other issues.[1]

1 For more information see the following websites:

Foodsafety.gov; Recalls https://www.fda.gov/safety/recalls-market-withdrawals-safety-alerts and https://www.foodsafetynews.com/tag/2022-recalls/

Dietary supplements: https://www.fda.gov/consumers/consumer-updates/fda-101-dietary-supplements#:~:text=Dietary%20supplement%20labels%20are%20required,per%20serving%20of%20those%20ingredients

Poison control centers; https://www.poison.org/

MedWatch, the safety reporting portal; https://www.accessdata.fda.gov/scripts/medwatch/index.cfm?action=reporting.

There are many public health surveillance systems in place to protect the public. Some surveillance systems are used to identify new or monitor health issues (outbreaks or disease tracking) (syndromic surveillance) [29]. There are active and passive systems, as well as adverse drug events (ADE) surveillance [30]. With all this infrastructure, we should feel safe; however, new industries arise and the infrastructure needs to catch up. For example, the 2018 Hemp Farm Bill aka The Agriculture Improvement Act of 2018 [31] (2018 Farm Bill) authorized the production of hemp and removed hemp and hemp seeds from the Drug Enforcement Administration's (DEA) schedule of Controlled Substances. It also directed the US Department of Agriculture (USDA) to issue regulations and guidance to implement a program to create a consistent regulatory framework around the production of hemp throughout the United States. The establishment of hemp as a regulated commodity also paved the way for US hemp farmers to participate in other USDA farm programs. What happened?

The hemp loophole allows marketers and manufacturers of unregulated hemp-derived synthetic products to capitalize on the public's ignorance or counting on the "feeling safe" perception. While the public has a false sense of safety as well as access to a product that is marketed as "free weed that gets you high." The term "free" is correct, free from regulation, free from lab testing, and free from all the requirements that a safe product (medical/regulated *Cannabis*) must demonstrate. All agencies are currently trying to address the issue of hemp and *Cannabis*. Their mission is to keep the public safe and healthy. The current times are reminiscent of a period greater than 100 years ago as well as within the past 25 years. This time around, the hemp industry is pushing synthetically derived isomers of hemp without regard for the public's health. At the same time, some companies in the *Cannabis* industry produce reproducible products with labeling and follow GMPs, so it appears like these products could be registered with the FDA. The regulated *Cannabis* industry may have the opportunity to override the hemp industry. The hemp industry should implement the same standards required by the regulated state *Cannabis* market laws, which would include standardized testing and clear product labels. The goal would be to regulate the synthetically derived hemp products on the market, removing products that do not follow regulations or regulatory guidelines. The public would then be safer if both *Cannabis* and hemp products followed the same regulations. A clear labeling of the amount of total THC as well as the naturally occurring amounts of the individual THC components would allow the consumer to have transparency to the product they were purchasing. In addition, the US Poison Control Center phone number as well as a QR code to report any effects the consumer experiences when using these products should be included on the packaging.

Other considerations that could be included come from American Public Health Association (APHA) public health approach to commercially legalized *Cannabis* [32] outline packaging recommendations that disclose all ingredients, including flavoring agents (and sources, e.g., *Cannabis*-derived, botanically derived, or synthetic) and diluents, as well as the percentage and milligrams of THC and CBD. Products should display a surgeon general's warning about specific health risks [33, 34]. Furthermore, warnings should address all key conditions with associated harm [35]. A proposal for a universal symbol on the product packaging (layer closest to the product) denoting that the product contains *Cannabis* as well as child-resistant packaging [36] and what is not working with education [37].

The mantra of "efficacy gets you to the market, safety keeps you on the market" can be shortened to "safety keeps you on the market." We would be moving the feeling safe needle to being safe skirting another public health threat.

The call for education and documenting your experience

One of the main ingredients in keeping the public safe is information. How do we inform? Through education. Education is key since the early days; we are taught and learn by trial and error. There are risks to this method, hence the use of randomized clinical trials; however it would be impossible to study everything. As a society, we educate through communication. When my sons were young, I would walk them through the garden and tell them what they could eat and what they could not eat and the importance of recognizing the difference. Understand what you are ingesting: if it is not in a package with ingredients listed on it, think twice! All these concepts remain true for anything we ingest. If something does not look or smell right, don't eat it.

Education is required for the public to understand what they are consuming. This includes regulated food labels, for example. Most recently, the FDA issued a request for information [38] about sugars like D-tagatose and isomaltulose, which are metabolized differently than traditional sugars, e.g., table sugar. These sweeteners meet the definition of sugar but do not have the same effects in the body as sucrose. The change resulted in a subheader to the sugar line on a food label called "added sugars" [39, 40], as well as a warning label for products that contain aspartame a sweetener that contains phenylalanine – high levels of which has the potential to cause brain damage.

Let's demand better labeling for hemp and *Cannabis* products. The consumer should be entitled to know what they are consuming. Corroon, MacKay, and Dolphin (2020) highlight that untruthful and unsubstantiated claims on CBD products cause more harm than good to the public's health [41]. Let's stick with evidence-based medicine. Remember, we, the people, are the guinea pigs for certain industries. You can become an advocate for better labeling and participate in informing the public by reporting unwanted events or adverse/unexpected effects, when you experience them or are aware of them. Use the free tools in this chapter to improve the understanding of products and inform the public's health (see Figures 2 and 3).

Figure 2: A QR Code That Links to an Adverse Event Reporting Tool.

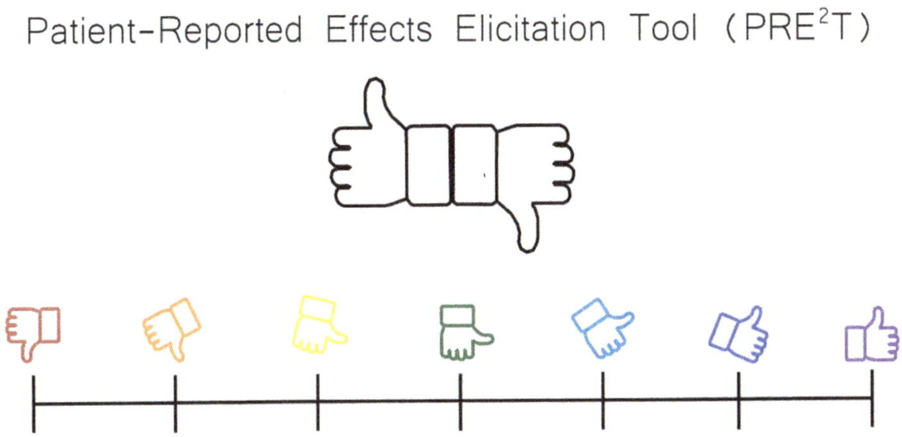

Red = unbearable, severely impacted my daily activities, had to stop taking
Orange = very distressing, many change to my daily activities
Yellow = discomforting but tolerable, changes to a few of my daily activities
Dark Green = NO side effect; no change to daily activities
Blue = notice a change; a good feeling; changes to a few of my daily activities
Dark Blue = notice a pleasant feeling; many change to my daily activities
Violet = a feeling of euphoria; very pleasant; a wanting of more; impacted my daily activities

Figure 3: A Patient/Consumer Reported Effects Elicitation Tool.

In summary, how do we keep a product on the shelf? There are processes in place. Manufacturers should follow the appropriate guidelines. Consumers should use products with complete package labeling including where the product is made and a list of ingredients. If you have an adverse outcome, no matter how minor, report it to CAERS, FAERS, or www.Budsinfo.com. A surveillance program like DAWN could be implemented to receive reports of adverse events caused by different forms of *Cannabis* consumption and potentially intervene when contaminated products or dangerous doses are consumed [42].

Resources

Given the paucity of monitoring and surveillance systems specifically for *Cannabis* [43], there is a request for an increased data collection through survey and surveillance tools. There are two tools that have been developed and are available for use. One is a QR code to report any unwanted effect a consumer may experience with a *Cannabis*, hemp, or synthetically derived product (see Figure 2) [23].

Another tool (Figure 3) is a patient- or participant-reported effects elicitation tool (Pre^{2}t). This tool is used to assess the range of effects a consumer may experience. It is designed to understand the level of intensity of a person's effect or reaction to a product [44]. This includes the desire for wanting more, thereby assessing abuse and addiction potential. Both tools are available for use and were designed for researchers to use. Lastly is the call for education, for healthcare providers (HCP) and especially for adolescents. The HCP program is available to all at https://www.prc-trials-plus.com/education. There is a pilot program in development called the *Cannabis* Awareness Program (CAP) that was presented at the APHA 2022 annual meeting [45]. For details on these education programs, write to prctrials.info@gmail.com or visit https://www.prc-trials-plus.com/education.

References

[1] Johnson S. The Ghost Map: The Story of London's Most Terrifying Epidemic – And How It Changed Science, Cities, and the Modern World. London England: penguin group penguin books Ltd, 2006.
[2] Sinclair U. The Jungle. Upton Sinclair, 1920.
[3] Gordon A, DeVlieger D, Vasan A, Bedard B. Chapter 1 – Technical considerations for the implementation of food safety and quality systems in developing countries, Editor: Gordon A Food Safety and Quality Systems in Developing Countries. Academic Press, 2020; pp. 1–40. Pages https://doi.org/10.1016/B978-0-12-814272-1.00001-2. ISBN 9780128142721.
[4] How the Food Industry Defines Botanicals: Food Processing Newsletter; June 4, 2012; accessed January 29, 2023. https://www.foodprocessing.com/product-development/rd/article/11362705/how-the-food-industry-defines-botanicals
[5] Center for Drug Evaluation and Research (CDER); What is a botanical Drug? Accessed January 29, 2023. https://www.fda.gov/about-fda/center-drug-evaluation-and-research-cder/what-botanical-drug
[6] Miller MT. Thalidomide Embryopathy: A Model for the Study of Congenital Incomitant Horizontal Strabismus. Trans Am Ophthalmol Soc 1991;81:623–674. PMC 1298636. PMID 1808819.
[7] Loue S, Sajatovic M. Encyclopedia of Women's Health. Springer Science & Business Media, 2004; 643–644. ISBN 9780306480737.
[8] Sneader W. Drug Discovery: A History. Rev. and updated ed. Chichester: Wiley, 2005; 367. ISBN 978-0-471-89979-2.
[9] Cuthbert A. The Oxford Companion to the Body. Oxford University Press, 2003; 682. doi: 10.1093/acref/9780198524038.001.0001. ISBN 9780198524038.
[10] Vargesson N. Thalidomide-induced Teratogenesis: History and Mechanisms. Birth Defects Res C Embryo Today Rev 2015;105(2):140–156. doi: 10.1002/bdrc.21096.

[11] Bren L (28 February 2001). "Frances Oldham Kelsey: FDA Medical Reviewer Leaves Her Mark on History". FDA Consumer. U.S. Food and Drug Administration. Retrieved 23 December 2009.

[12] Palamar J. How Ephedrine Escaped Regulation in the United States: A Historical Review of Misuse and Associated Policy. Health Policy 2011;99(1):1–9. doi: 10.1016/j.healthpol.2010.07.007.

[13] Zell-Kanter M, Quigley MA, Leikin JB. Reduction in Ephedra Poisonings after FDA Ban. N Engl J Med 2015;372(22):2172–2174. doi: 10.1056/nejmc1502505.

[14] Bonini SA, Premoli M, Tambaro S, et al. Cannabis Sativa: A Comprehensive Ethnopharmacological Review of A Medicinal Plant with A Long History. J Ethnopharmacol 2018 Dec 5;227:300–315. doi: 10.1016/j.jep.2018.09.004. Epub 2018 Sep 8. PMID: 30205181.

[15] Charitos IA, Gagliano-Candela R, Santacroce L, Bottalico L. The Cannabis Spread Throughout the Continents and Its Therapeutic Use in History. Endocr Metab Immune Disord Drug Targets 2021;21 (3):407–417. doi: 10.2174/1871530320666200520095900. PMID: 32433013.

[16] Biswas P, Mishra P, Bose D, Durgbanshi A. Cannabis: A Neurological Remedy or A Drug of Abuse in India. CNS Neurol Disord Drug Targets 2017;16(5):576–584. doi: 10.2174/1871527316666170424115008. PMID: 28440191.

[17] McBride JR, Cavero RY, Cheshire AL, Calvo MI, McBride DL. Exchange of Medicinal Plant Information in California Missions. J Ethnobiol Ethnomed 2020;16:35.

[18] Gieringer DH. The Forgotten Origins of Cannabis Prohibition in California. Contemp Drug Probl 1999;26(2):237–288. doi: 10.1177/009145099902600204.

[19] Deitch R. Hemp – American History Revisited. New York City: Algora Publishing, 2003; 16. ISBN 978-0-87586-206-4.

[20] Frontline PBS "Busted – America's War on Marijuana". pbs.org. Retrieved September 4, 2014; accessed January 29, 2023: https://www.pbs.org/wgbh/pages/frontline/shows/dope/etc/cron.html

[21] Swenson B Hemp and Flax in Colonial America CW Journal: Winter 2015. https://research.colonialwilliamsburg.org/foundation/journal/Winter15/hemp.cfm

[22] Johnson L CBD Oracle Lab Study Shows Some Delta-8 Products are 7700% over the Legal Delta-9 THC Limit. ORACLE. Available from: https://cbdoracle.com/news/delta-8-thc-products-market-study-consumer-safety-and-legality/#The-labeling-lies. Accessed January 18, 2023.

[23] Simon TA, Simon JH, Heaning EG, Gomez-Caminero A, Marcu JP. Delta-8, a Cannabis-Derived Tetrahydrocannabinol Isomer: Evaluating Case Report Data in the Food and Drug Administration Adverse Event Reporting System (FAERS) Database. Drug Healthc Patient Saf 2023 Jan 29;15:25–38. doi: 10.2147/DHPS.S391857. PMID: 36742440; PMCID: PMC9894081.

[24] Knopf A. CDC and FDA Warn of Delta-8 THC Harms. Alcohol Drug Abus Wkly 2021;33(36):7–7. doi: 10.1002/adaw.33199.

[25] FDA. 5 Things to Know about Delta-8 Tetrahydrocannabinol – Delta-8 THC. https://www.fda.gov/consumers/consumer-updates/5-things-know-about-delta-8-tetrahydrocannabinol-delta-8-thc

[26] Simon T, Silverstone A, Shaw Y, Marcu J. "Free Weed that Gets You High": Did the Fda and Cdc Health Alerts on DELTA-8 Thc Reach Mainstream Social Media? Research Triangle Park, NC, USA: International Cannabinoid Research Society, 336. 2022.

[27] Substance Abuse and Mental Health Services Administration. (2022). Key substance use and mental health indicators in the United States: Results from the 2021 National Survey on Drug Use and Health (HHS Publication No. PEP22-07-01-005, NSDUH Series H-57). Center for Behavioral Health Statistics and Quality, Substance Abuse and Mental Health Services Administration. https://www.samhsa.gov/data/report/2021-nsduh-annual-national-report

[28] Psychological Perspectives on Perceived Safety. Social Factors of Feeling Safe: A Multidisciplinary Perspective, Perceived Safety. January 2019; pp.43–60. doi: 10.1007/978-3-030-11456-5_4.

[29] National Syndromic Surveillance Program. What is syndromic surveillance? Available at: https://www.cdc.gov/nssp/overview.html. Accessed February 20, 2023.

[30] The Office of Disease Prevention and Health Promotion: National Action Plan for Adverse Drug Event Prevention. Section 2 Surveillance Resources. US Department of Health and Human Services. 2014.
[31] Hemp The Agriculture Improvement Act of 2018. U.S. Department of Agriculture accessed February 5, 2023. https://www.usda.gov/topics/hemp
[32] A Public Health Approach to Regulating Commercially Legalized Cannabis. Oct 24, 2020; Policy Number: 20206. https://www.apha.org/policies-and-advocacy/public-health-policy-statements/policy-database/2021/01/13/a-public-health-approach-to-regulating-commercially-legalized-cannabis
[33] Alcoholic Beverage Labeling Act of 1988, 27 USC §§ 213 et seq.
[34] Alcoholic Beverage Health Warning Statement, 27 CFR Part 16.
[35] National Academies of Sciences, Engineering, and Medicine. The Health Effects of Cannabis and Cannabinoids: The Current State of Evidence and Recommendations for Research. Washington, DC: National Academies Press, 2017.
[36] Volkow ND, Baler R. Emergency Department Visits from Edible versus Inhalable Cannabis. Ann Intern Med 2019;170(8):569–570.
[37] Westley E, Miller M, Goodvin R, Hicks C (2019). Updated inventory of programs for the prevention and treatment of youth cannabis use (document 19-12-3201). Available at: http://www.wsipp.wa.gov/ReportFile/1716/Wsipp_Updated-Inventory-of-Programs-for-the-Prevention-and-Treatment-of-Youth-Cannabis-Use_Report.pdf. Accessed July 6, 2020
[38] FDA Seeks Input on Nutrition Labeling for Certain Sugars and Issues Final Guidance on Allulose. https://www.fda.gov/food/cfsan-constituent-updates/fda-seeks-input-nutrition-labeling-certain-sugars-and-issues-final-guidance-allulose; Accessed January 29, 2023.
[39] FDA in Brief: FDA Looking at Nutrition Facts Labeling of Certain Sugars, Sweetener. https://www.fda.gov/news-events/fda-brief/fda-brief-fda-looking-nutrition-facts-labeling-certain-sugars-sweeteners Accessed January 29, 2023.
[40] Added Sugars on the New Nutrition Facts Label. https://www.fda.gov/food/new-nutrition-facts-label/added-sugars-new-nutrition-facts-label. Accessed January 29, 2023.
[41] Corroon J, MacKay D, Dolphin W. Labeling of Cannabidiol Products: A Public Health Perspective. Cannabis Cannabinoid Res 2020 Dec 15;5(4):274–278. doi: 10.1089/can.2019.0101. PMID: 33381640; PMCID: PMC7759277.
[42] Drug Abuse Warning Network (DAWN) SAMHSA; accessed February 20, 2023. https://www.samhsa.gov/data/data-we-collect/dawn-drug-abuse-warning-network
[43] Geissler KH, Kaizer K, Johnson JK, Doonan SM, Whitehill JM. Evaluation of Availability of Survey Data about Cannabis Use. JAMA Netw Open 2020 Jun 1;3(6):e206039. doi: 10.1001/jamanetworkopen.2020.6039. PMID: 32520358; PMCID: PMC7287570.
[44] Simon TA, Khouri MS, Kou TD, Gomez-Caminero A. Realizing the Potential of the Patient Perspective. Patient Prefer Adherence 2020 Oct 22;14:2001–2007. doi: 10.2147/PPA.S257355. PMID: 33132698; PMCID: PMC7591823.
[45] Wabwire M, Simon TA, Marcu J, Smale T, Ward SJ Get Your Thinking Cap On; Implementation of a Cannabis Awareness Program for High School Students; #5047; American Public Health Association Annual Meeting Boston Nov. 2022.

Index

https://doi.org/10.1515/9783110677508-010

www.ingramcontent.com/pod-product-compliance
Lightning Source LLC
Jackson TN
JSHW061327270326
99861JS00030B/767